HARRY PELISS
ETONHURST

CAMBRIDGE
IGCSE Geography
A Complete Guide 2020 - 2022

Note from the Author

As a Passionate Geographer, I was inspired to write the first edition of this guide to share my interests in such a varied subject and to aid studies in the latest Cambridge IGCSE Geography specification. I am delighted to introduce the second edition, updated for the new specification, along with a new cleaner design to better aid your learning.

From explosive volcanoes, to the sheer beauty of both human and physical landforms — our planet's Geography has the power to light that inquisitive spark. For me, this is why I love Geography. The world is open, ready to be explored and discovered — jammed with insight and wonder over both the rocks beneath our feet and the glistening megapolis that spreads out as far as the eye can see.

From me, best of luck in your studies; I hope you discover an element of Geography that innately excites you.

Zachary Elliott

igcsegeoguide.com

CAMBRIDGE
IGCSE Geography
A Complete Guide 2020 - 2022

Second Edition

Written and Produced by **Zachary Elliott**

For the 2020-22 Cambridge IGCSE Geography 0460 specification

Cambridge IGCSE Geography – A Complete Guide

2nd Edition – Updated for the 2020 – 2022 Syllabus

Copyright © **2018 Zachary Elliott**
Published 2018
All rights reserved.

No parts of this publication may be reproduced, stored in a retrieval system, or transmitted in any form or by any means, electronic, mechanical, photocopying, recording, or otherwise, without the prior written permission of the copyright owner.

This book is sold subject to the condition that it shall not, by way of trade or otherwise, be lent, resold, hired out, or otherwise circulated without the publisher's prior consent in any form of binding or cover other than that in which it is published and without a similar condition including this condition being imposed on the subsequent purchaser. Under no circumstances may any part of this book be photocopied for resale.

The subject content information given in this guide is taken from the Cambridge IGCSE Geography 0460 syllabus for examination from 2020 – 2022. Please ensure that you refer to the full details included in the syllabus document for the year you are entering for examination, available on their public website at: "http://www.cambridgeinternational.org/programmes-and-qualifications/cambridge-igcse-geography-0460/"

Please note that all internet links in this guide were correct at time of publishing and may have broken or changed over time. The author recommends searching for the page's new location if the published link no longer points to the webpage.

Many of the photographs within this guide are from Creative Commons (CC). Please locate the credits in the Acknowledgement section to the rear of this guide (which correspond to Image or Figure numbers in the respective captions), and the license links in the "Licenses for Freely Available Media Used" section. The author will be happy to address any concerns over usage of content as quickly as possible and will do their best to make any amends. The author would like to thank all authors of their respective content for their work and for making it publicly available under open licenses.

All other photographs © **2018 Zachary Elliott**

Cover Photography © 2018 Zachary Elliott – Mount Washington, OR, USA

CONTENTS

How to Use This Guide 2

Theme 1: Population and Settlement 4

Population 5

    1.1 Population dynamics 6

    1.2 Migration 18

    1.3 Population Structure 22

    1.4 Population Density and Distribution 27

Settlement 31

    1.5 Settlements and Service Provision 32

    1.6 Urban Settlements 40

    1.7 Urbanisation 48

Theme 2: The Natural Environment 56

    2.1 Earthquakes and Volcanoes 57

    2.2 Rivers 75

    2.3 Coasts 91

    2.4 Weather 105

    2.5 Climate and Natural Vegetation 115

Theme 3: Economic Development 134

    3.1 Development 135

    3.2 Food Production 143

    3.3 Industry 151

    3.4 Tourism 159

    3.5 Energy 167

    3.6 Water 175

    3.7 Environmental Risks of Economic Development 183

Licences for Media Used 195

Acknowledgments 196

Command Words 198

Glossary 199

Complete Contents 212

HOW TO USE THIS GUIDE

Welcome to IGCSE Geography: A Complete Guide.

This guide is divided into sections that follow the official CIE syllabus, so you can more easily understand what you will be assessed on.

At the **start** of each unit or section, there are references to the syllabus' subject content information. This is so you can understand what you need to know in that area. This is formatted in light blue.

Throughout, there are diagrams to complement your learning and to enhance your understanding.

Located at the **rear** of the guide is a glossary to help you understand key geographical terms, and to provide reference during learning.

Abbreviations

MEDCs = More Economically Developed Countries
LEDCs = Less Economically Developed Countries
NICs = Newly industrialised Countries

Guide to colour

Key Terms / Key Points

Subject Content Information

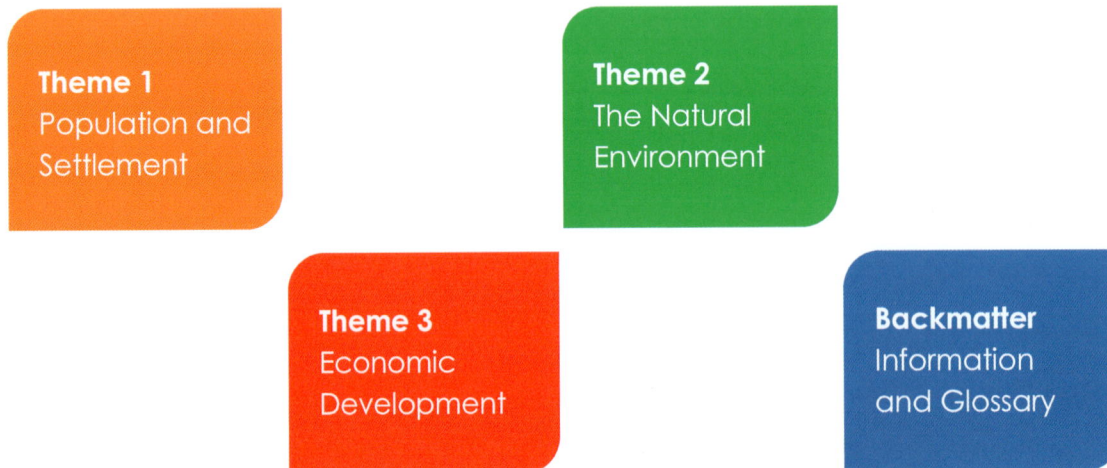

HOW TO USE THIS GUIDE

POPULATION AND SETTLEMENT

THEME 1: POPULATION AND SETTLEMENT

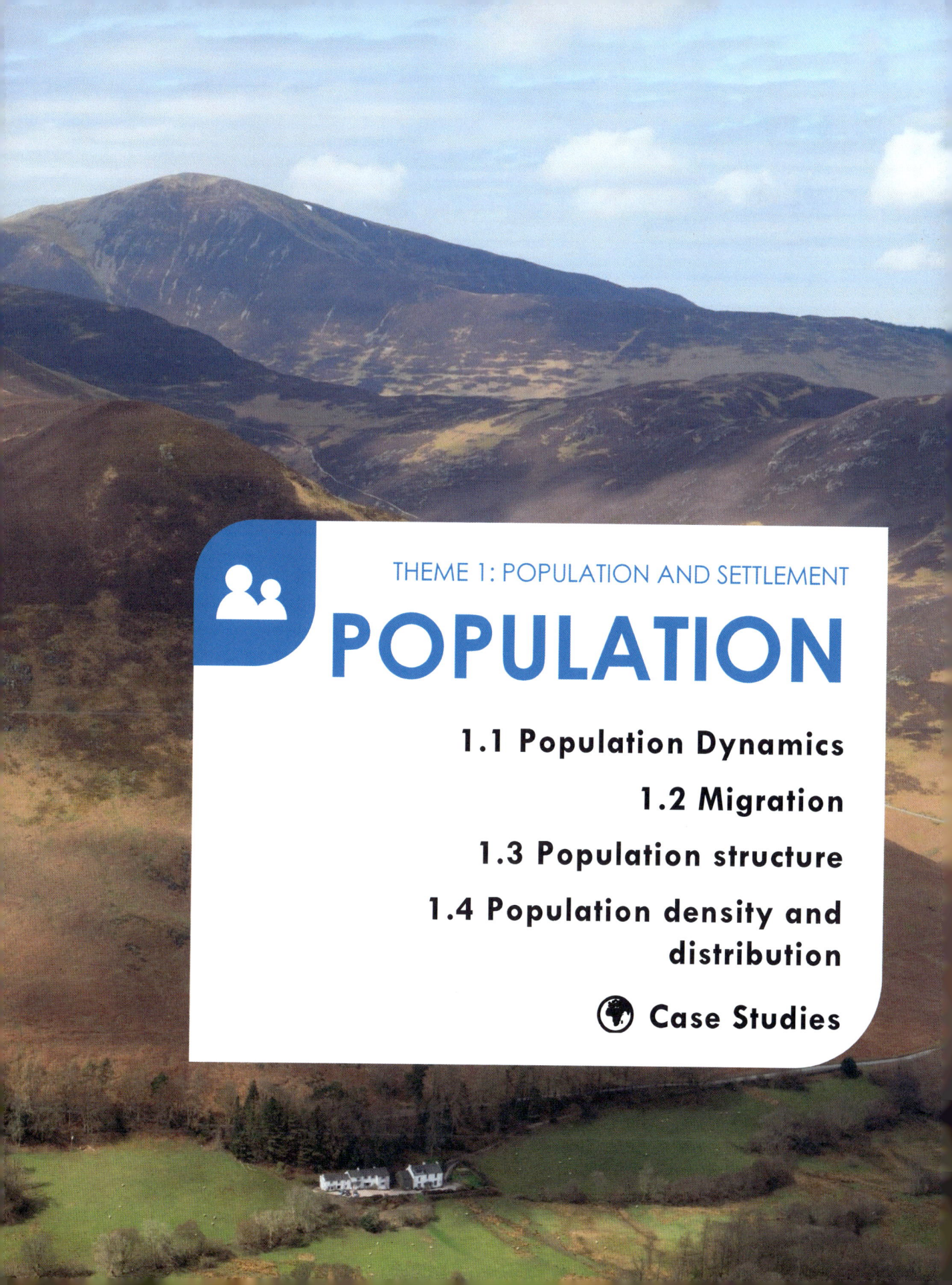

THEME 1: POPULATION AND SETTLEMENT

POPULATION

1.1 Population Dynamics
1.2 Migration
1.3 Population structure
1.4 Population density and distribution

🌍 Case Studies

THEME 1
POPULATION AND SETTLEMENT

1.1 POPULATION DYNAMICS

Describe and give reasons for the rapid increase in the world's population

Show an understanding of over-population and under-population

Understand the main causes of a change in population size

Give reasons for contrasting rates of natural population change

Describe and evaluate population policies

Reasons for a rapid increase in the world's population

Describe and give reasons for the rapid increase in the world's population

Pre-growth (pre-1700s)

Before the boom, population growth was slow and steady with few growth fluctuations. Most countries were in Stage 1 of the Demographic Transition Model – i.e. mainly composed of remote tribes (*see later section for more on this*). There were low life expectancies, large families and consequently few surviving people (as people died young). These factors contributed to a low total population and population growth rate before industrialisation.

Some factors that contributed to this included:

- Limited access to reliable food supplies
- Lack of clean water
- Very limited health care
- Poor shelter
- Unhygienic living environments

Later there were improvements in machinery due to the Agricultural and Industrial revolutions, meaning that more people were free to work in urban areas. More Countries entered Stage 2 of the Demographic Transition Model.

1800s (19th Century)

In the 1800s countries became more prosperous – meaning they were now able to support larger populations. Having more than 10 children became common and, due to an increased birth rate, there was greater population growth. Death rates then started to drop due to new medical advances and greatly improved sanitation.

Additionally, in Europe, there was increased availability of clean water and proper sanitation at the beginning of the 20th Century (1900s).

These factors lead to a dramatic boom in population at the beginning of the 1900s, partly due to the rapidly declining death rate.

1.1 POPULATION DYNAMICS

1900s (20th Century) - LEDCs

However, LEDCs were still far less developed than MEDCs – illustrated by very high birth-rates and family sizes (being much earlier on in their economic development).

These LEDCs still maintained benefits from having many children – including a greater chance of continuance of the family surname (via a boy). In addition, they had higher birth rates than many MEDCs because of widespread polygamy in many African states coupled with Islamic views against contraception.

1900s (20th Century) - MEDCs

In 1939, birth rates began to fall in MEDCs. This was partly a result of many women in MEDCs joining the workforce after WW2, meaning they were less likely to have children (thus, the birth rate decreases as less women had babies).

Even though people's life expectancies in more developed countries (e.g. DTM 4) were increasing due to improved medical care, the total population of countries like Italy and Sweden began to fall. This was because of changing attitudes towards child birth (DTM Stage 5).

People began to realise that they could obtain better standards of living if they had to support less children. In some countries, including the UK, many women are delaying having children until they are 40, so they can have a successful career – further reducing birth rates.

The global population is set to level within the next century. This is because many LEDCs will become more developed coming into the 22nd Century (2100s), due to reduced birth rates – women wanting to work – along with improved sanitation and access to clean water.

This could be a result of political shifts, international aid, financial injections, or even government incentives to increase economic potential (more healthy people means a larger workforce, resulting in more people paying taxes).

THEME 1
POPULATION AND SETTLEMENT

Overpopulation and Underpopulation

Show an understanding of over-population and under-population

Causes and consequences of over-population and under-population

Overpopulation

Where there are too many people to be supported to a good standard of living by the resources of the country

Underpopulation

When countries have insufficient workers to exploit their resources efficiently, to support retired populations and to provide growth.

i.e. too few people to use all the resources of a country to the maximum efficiency.

Problems Caused

Overpopulation	Underpopulation
• Unemployment • Shortage of services • Shortage of houses – informal settlements increase • More congestion • Inflation (rising prices) • Increased pollution	• Shortage of workers • Closing of services • Less development / innovation • Less people to pay tax • Derelict cities / towns • Increased tax rates • Wasted food / resources

NB   Optimum population

is when there are enough resources to support the population to a good standard of living.

Main causes of a change in population size

Understand the main causes of a change in population size

How birth rate, death rate and migration contribute to the population of a country increasing or declining

There are 3 main causes of population change:

Births	Measured using the birth rate **(number of live births per 1000)**
Deaths	Measured using the **death rate** (number of deaths per 1000)
Migration	The movement of people in and out of an area (the difference between the 2 movements is the **net migration rate**)

1.1 POPULATION DYNAMICS

Life Expectancy is the average age that someone is expected to live in a certain country

Infant Mortality is the number of new-born deaths under the age of 1 per 1000 live births

Reasons why there could be a:

	Birth Rate	Death Rate
High	• Lack of contraception • Religious beliefs (e.g. against contraception or for polygamy) • High infant mortality	• Natural Disasters • Conflicts • Poor health • Shortage of clean water • Diseases (e.g. HIV, AIDS, Polio, Malaria)
Low	• Availability of contraception • Improved education • Reduced infant mortality • Cost of raising children	• Immunisation • Improved health • Availability of clean water

Effects:

	Birth Rate	Death Rate	Migration Rate
High	• More people are born • There are more young people	• More people die • There are less old people	• More people become new citizens
Low	• Less people are born • There are less young people	• Less people die • There are more people for longer	• Less people become new citizens

Contrasting Rates of natural population change

Give reasons for contrasting rates of natural population change

Impacts of social, economic and other factors (including government policies, HIV/AIDS) on birth and death rates

Births and deaths are natural causes of population change. The difference between the 2 rates is called the natural increase / decrease.

Natural increase: When the **birth rate** is **higher** than the **death rate** (DTM 2+3)
Natural decrease: Where the **death rate** is **higher** than the **birth rate** (DTM 5)

Natural Increase = birth rate (per 1000) – death rate (per 1000)

Population growth rate (%) = Natural Increase / 10

9

THEME 1
POPULATION AND SETTLEMENT

Creating a Natural Increase: Socio-Economic Causes of...

Rising Birth Rates	Falling Death Rates
• To carry on family name • To ensure care in old age • Polygamous Societies • Religious Views against contraception • Desire to have more children to work in family business / farm	• Medical Knowledge • Improved Sanitation • Economic Support (e.g. Pensions) • Free Healthcare

Other population change factors:

Factor	Government Policies	HIV / AIDS
Effect	**Birth Control:** Reduces Birth Rates **Free Healthcare:** Reduces Death Rates	**Death Rate** increases **Life Expectancy** decreases **Infant Mortality** increases
Pop. Change	Neutral Population Change	Population Decrease

Demographic Transition Model
Also referred to as the DTM

The transition from a country having high birth and death rates to having low birth and death rates over time. This allows it to show how changes in the country as it develops (the birth and death rates becoming lower).

This Model is divided into 5 stages. Stage 1 is poor and undeveloped (e.g. a remote tribe); whereas Stage 5 is rich and very developed (e.g. the most developed countries, like Germany).

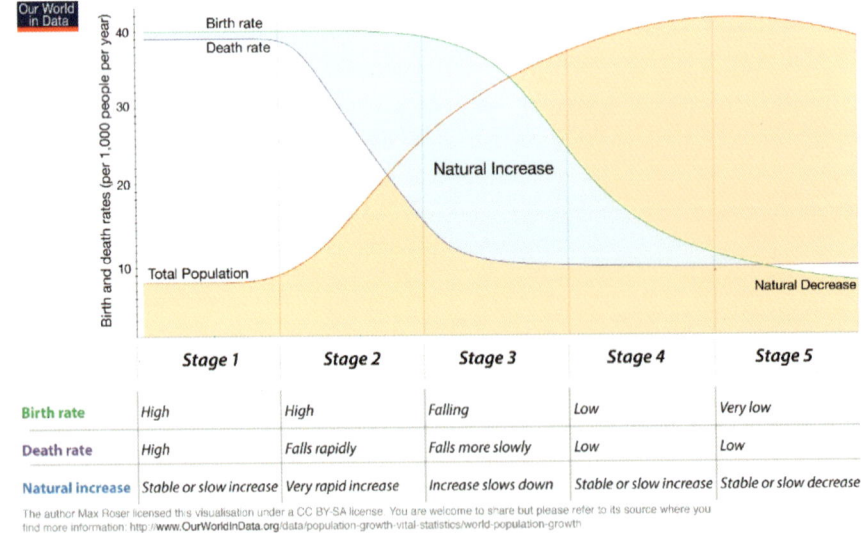

Figure 1: Diagram of the Demographic Transition Model (DTM)

1.1 POPULATION DYNAMICS

Demographic Transition Model Stage Comparison

DTM Stage	Death Rate	Birth Rate	This means	Places
Stage 1	**High** because of disease, famine, lack of clean water and lack of medical care	**High** due to lack of birth control; women marry young as children are needed for work	The **Natural Increase** is **Low**	Remote isolated tribes
Stage 2	**Starting to fall** because of better medicine, water, food and sanitation	**Still high**, due to lack of birth control; women marry young	**Natural Increase** is **High**	LEDCs
Stage 3	**Still falling** because of better medicine, water, food and sanitation	**Falling** as fewer children are needed for work, birth control exists, infant deaths are falling and education is better	**Lower Natural Increase**	NICs
Stage 4	**Remains Low**	**Low** – because of birth control, people can choose how many children they want	There is little or no natural increase	MEDCs
Stage 5	**Goes up slightly** because more people are old	**Remains low** and can fall below the death rate; changes in lifestyle mean people have fewer children	**Natural Decrease** (dependant on migration)	A **few MEDCs** like Japan, Italy and Germany

THEME 1
POPULATION AND SETTLEMENT

Population Policies – China's 1 Child Policy

Describe and evaluate population policies

NB *This is an **anti-natalist** policy. A **pro-natalist** policy encourages couples to have more children*

Many countries have introduced population policies to either promote or reduce population growth.

This is normally done by encouraging people to have more or less children.

China's 1 child policy is one of the best known.

Background

- China's population grew rapidly during the 1950s and 1960s
- China didn't have enough resources to support this population
- So, in 1979, the government introduced the 1 child policy to help to limit population growth
- Financial and welfare benefits to couples with only 1 child provided an incentive to follow the policy
- Fining was used as a penalty for parents with more than 1 child. (Apparently, sterilisations and forced abortions were also used)
- The policy has prevented hundreds of millions of babies from being born, so China's population is lower than it would have otherwise been

Impacts

Social

Because of the family-centred Chinese culture, it is rare to find retirement homes. This is because much of the elderly live with their families. Like many places in the world, most Chinese children will have to care for 4 elderly grandparents and 2 elderly parents throughout their life.

However, there are not enough young people to look after the elderly – meaning they must go to old people's homes. There are too many elderly people to fit in the retirement homes in China, so more will need to be built as a result.

Chinese culture tends to favour boys. As the couples can only have 1 child, baby girls are more likely to be abandoned. This means that men will begin to massively outnumber women, leading to social conflict and more difficulty in men finding a partner.

Economic

The increasing aging population will need financial support in old age. This could include a pension or medical care. It is predicted that from 2025, China will have more elderly people than children. Consequently, there will be more

1.1 POPULATION DYNAMICS

people over 65 than people of working age as time progresses: the long-term effect of the 1 child policy.

Because of this, China's growing economy and businesses won't have enough people for the required workforce in the future – so expansion and growth will be stunted.

Less people are starting work, so there are less workers to contribute to the economy – meaning that the economy will struggle to support the increasing elderly populations. This cycle continues, as the elderly population grows, and the youth population shrinks over time.

Additionally, this means that there is a reduced tax-paying working population. This means that taxes may increase – this is so the government has enough funds to provide public services for the population.

Positive

The policy has reduced the fertility rate – meaning that the policy was effective in the respect that it achieved the original aim in reducing future strain on China's resources, by limiting population growth.

Also, overpopulation has been reduced in many of the crowded South-East regions of China – relieving pressure on housing and services.

Future of the policy

The policy began to be formally phased out in 2015 due to the negative impacts it created.

The major reason for the end of the original policy was due to fears over labour shortages.

Image 1: For a prosperous, powerful nation and a happy family, please practice family planning

THEME 1
POPULATION AND SETTLEMENT

Case Studies

- **A country which is over-populated**
- **A country which is under-populated**
- **A country with a high rate of natural population growth**
- **A country with a low rate of population growth (or population decline)**

Overpopulation – Bangladesh

Bangladesh is frequently flooded and affected by tropical storms.

There are few natural resources. The country also relies on farming as an industry – with over 40% of the population working in this primary industry.

Minerals, which are domestically available, are used in local manufacturing industries.

Bangladesh has the 8th largest population in the world (as of July 2017) – of over 157 million people. However, it only has the 92nd largest land area. This means it has a high population density of over 1000 people per km².

Religion contributes to this large population size. In Bangladesh, Muslims make up some 89% of the population (2013). Some Muslim leaders here don't support the use of contraception. This means that the birth rate is higher, creating a high population growth rate.

Image 2: Flooding in Bangladesh

With a high birth rate of 18.8 per 1000 (2017) and death rate of 5.4 per 1000 (2017) – the resulting population growth rate is 1.04% - meaning Bangladesh has more people than its resources can support. This has resulted in overpopulation.

The GDP per person is too low to support a good standard of living for each citizen. Many people are under-employed. People that are employed often don't work much, earning little money. The exports sometimes don't create high profits either.

In addition, there aren't enough public services for the population. Less than 75% of people in Bangladesh are literate – a result of no real compulsory or free education system. This means that many people have weak or no qualifications. The access to health care is also poor, raising the death rate and infant mortality rate.

1.1 POPULATION DYNAMICS

Outlook: Bangladesh will struggle to support a larger population

Why?

- The farming land is overcultivated and won't be able to support further growth of farming industries (which are key to Bangladesh's economy)
- There has been deforestation near the Himalayas, increasing the risk of flooding
- Money will be needed to repair damage caused by flooding, which means that less money is invested into public services
- The capital, Dhaka, is congested, overcrowded and overpopulated

Underpopulation – Australia

With a low population (of 23 million) and a working age population of 15.3 million (2017), coupled with a massive land area of over 7.6 million km², Australia is under-populated – with around 3 people per km² of land.

As a country, Australia is rich in resources: containing large reserves of coal, gold, natural gas, metal ores and uranium. This uranium is crucial in permitting usage of nuclear power globally. To exploit these resources effectively, Australia needs to attract migrant workers, the migration rate being the 21th highest globally (2017).

There are more resources available than the country requires. This means that the surplus resources can be exported. In 2017, the exports were worth over $220 billion, with a gross domestic product (GDP) per capita of over $49,900 per person. While the tertiary sector employs about 70% of working Australians, the unemployment rate is also low.

Australia has low birth and death rates, which means there is a low rate of natural population increase. Australians also have excellent literacy (99% are literate), with education available for 21 years. The health care provisions are equally good, meaning there is a low infant mortality rate.

With its low population, there is land suitable for development around the desert / semi-desert areas, which could allow for future population increase. However, recent water shortages could inhibit the development in these areas.

Outlook: Australia could support a larger population

Image 3: Australian Outback

THEME 1
POPULATION AND SETTLEMENT

High rate of natural population growth – Bangladesh

Why is there a high rate of natural population growth?

- Muslims make up 85% of the population. Because some Muslim leaders here don't support the use of contraception, there is a higher subsequent birth rate
- Some people can't afford contraception, linked to a low GDP per capita
- Rarity of contraception and birth control measures
- Children are needed for work or to care for older family members
- Family Planning isn't available, meaning women have more children
- Poor health services mean that more people die. People can't afford the health services anyway, linked to a low GDP per capita. This means there is an increased Death Rate (due to famine, disease and lack of proper sanitation)

The GDP per capita is too low to support a good standard of living for each citizen. This may be because too many are under-employed or under-employed.

With a high birth rate of 18.8 per 1000 (2017) and death rate of 5.4 per 1000 (2017) – the resulting population growth rate is 1.04%. More people are born than those who die. This leads to Bangladesh's high rate of natural population growth.

Impact of high rate of natural population growth:

- Higher Demand for public services
- *Problems from Overpopulation*: including Overcrowding (like in the capital, Dhaka) and strain on Natural Resources
- Higher Job competition leading to potentially more unemployment
- Greater wealth segregation

Outlook: With the current growth rates, Bangladesh will have major issues with infrastructure, resources, overcrowding and unemployment in the future

Image 4: The Crowded Streets of Dhaka

1.1 POPULATION DYNAMICS

Declining population – Japan
DTM Stage 5

Why is there a declining population?

Less women are having babies - Changing attitudes to child birth (which reduces Japan's Birth Rate):

- Many women work in high-tech Japanese industries
- They choose to work, rather than taking time off to care for children
- Their careers may be affected by being a mother
- Fewer see the need to have children
- They could get a better standard of living if they have less children to support
- People are delaying having children, or choosing not to – like above

Improved medical care:

- Improves life expectancy and reduces death rate
- Potentially due to new medical technological advancements in Japan's tech industry

The population is declining because, even though there are medical improvements, more people are dying (often of old age) than are being born (because less women are having babies). This leads to a natural population decrease.

Consequences:

- A reduced working population
   » Because less people will be starting work, leading to a shortage of workers in the long-term
   » Reduced economic growth
   » Closing or reduction of some services
   » Less technological innovation – as there are less people to innovate in Japan's exciting technological sector - this reduces industrial development
- Higher dependency ratio
   » Less youths to support the retired population
   » A larger elderly population than youthful population
   » Increased taxes to pay for pensions and healthcare
- Less people to pay taxes, as there is a smaller workforce
- Derelict post-industrial towns because there are not enough workers to support some industries

Outlook: Japan needs to attract migrants or offer couples incentives for having a child if it wants to have a strong workforce into the future

THEME 1
POPULATION AND SETTLEMENT

1.2 MIGRATION

Explain and give reasons for population migration

Demonstrate an understanding of the impacts of migration

Key Words

Internal Migration	Within the country
External / International Migration	From one country to another
Voluntary Migration	People choose to move
Involuntary Migration	People are forced to move
Economic Migrants	People who choose to move for money (better jobs / wages)
Refugees	People migrating to escape from an event (e.g. war / famine)
Temporary Migration	Migrating for a short or set period (e.g. Seasonal Migrants)
Permanent Migration	People who move and don't go back

Reasons for Population Migration

Explain and give reasons for population migration

Internal movements such as rural-urban migration, as well as international migrations, both voluntary and involuntary

NB **Push Factors** make a migrant want to leave their origin country

Pull Factors make a migrant want to go to their destination country

Push Factors	Pull Factors
<ul><li>Not enough Jobs</li><li>Low wages</li><li>Poor housing</li><li>Poor Educational Opportunities</li><li>Poor Health Care</li><li>War with another country</li><li>Civil War / Lawlessness</li><li>Drought and Famine</li><li>Persecution</li><li>Natural Disasters</li><li>Cost of Living (e.g. in Hong Kong)</li></ul>	<ul><li>Hope of Finding a Job</li><li>Higher Wages</li><li>Chance of a Better Education</li><li>Better Health Care</li><li>A Better Standard of Living</li><li>Family and Friends in destination</li><li>Lower Crime Levels</li><li>Safety from Conflict</li></ul>

1.2 MIGRATION

Impacts of Migration

Demonstrate an understanding of the impacts of migration

Positive and negative impacts on the destination and origin of the migrants, and on the migrants themselves

Impacts on **Destination**

Positive	Negative
• More Low-Wage Workers • Boost to Local Economy • Job Fulfilment • Cultural Enrichment (e.g. Food) • Increased diversity	• Racial Tensions / Discrimination » Migrants 'Stealing Jobs' • Pressure on Natural Resources • Pressure on Public Services • Increased Job Competition • Overcrowding: Pollution

Impacts on **Origin**

Positive	Negative

• Reduced strain on resources
• Reduced Unemployment
• Reduced pressure on public services (e.g. schools)
• Migrants might return with new skills
	• Taxes will increase (there are now less people to tax)
	• Reduced size of workforce
	  » Less skilled workers – skilled migrants leave
	  » Reduced Economic Development
	• Separation of families by borders

Impacts on **Migrants**

Problems Faced – Intervening Obstacles	Positive impacts
• Running out of money • Language Barriers • Exploitation (Robbing/Kidnapping) • Passport, Visa or Immigration issues • Weather conditions • Problems with housing or accommodation when they get there • Illness, as there is often no available health care	• Better job on arrival • Escape from conflict • Better quality of life • A better education / job • A sense of integration / hope • Job opportunities

THEME 1
POPULATION AND SETTLEMENT

Case Study – Syrian Refugees' International Migration to Europe
An International Migration

Useful Terms

Persecution: When someone is prosecuted or attacked for a racist reason

Asylum Seekers are people trying to get refuge in a foreign country

Internally Displaced: When people are forced to domestically migrate (perhaps by conflict, natural disaster or famine)

Why are the Syrian Refugees migrating to Europe?

- Civil War
- Hope of a better, safer life
- Low quality of life
- The EU is increasingly becoming more open to asylum seekers

The civil war broke out in March 2011 and is ongoing. Since then, millions of Syrians have fled from their homes (or in some cases the country) from the war. They took refuge in neighbouring countries, like Turkey; attempted to seek asylum in the EU or were Internally Displaced inside Syria.

6.6 million are estimated to have been **Internally Displaced** out of the 11 million who fled from their homes.

The number of asylum seekers has massively increased in the past couple of years, especially in Germany, where the number of asylum applications has over doubled.

Image 5: Refugees arriving at Lesvos island

1.2 MIGRATION

Syrians' Push and Pull factors for going to Europe:

Push	Pull
• Syrian Civil War • Forced out of homes • Unemployed due to war	• Hope of a better life in Europe • Europe is safer • Job opportunities – Service Sector • Availability of public services – e.g. Education • Reliable food and water supplies • Proper shelter

Impacts of Syrian Migration to Europe:

	Origin	Destination	Migrants
Positive	• Reduced strain on resources • Reduction of citizen 'targets'	• Syrian Culture • Greater diversity • More low-wage Syrian workers	• A better quality of life • Safety – no civil war
Negative	• Loss of 'civil people' • Reduced workforce – leading to reduced tax income, and increased taxes	• Perhaps more discrimination • More pressure on services • More pressure on resources	• They may not speak the destination's language (frequently German) • Cultural integration issues • Exploitation

Many migrants are being financially exploited by people traffickers. They are part of the illegal (and sadly growing) business of transporting migrants across the Mediterranean Sea. This is done to force authorities into accepting asylum applications – which are accepted in an attempt to relieve some of the effects of the humanitarian crisis created by the Syrian civil war. These traffickers often have high fees – becoming wealthy by providing a low quality of transportation that is both overcrowded and dangerous for the refugees.

THEME 1
POPULATION AND SETTLEMENT

1.3 POPULATION STRUCTURE

Identify and give reasons for and implications of different types of population structure

Age/sex pyramids of countries at different levels of economic development

Useful Terms

The **Dependency Ratio** is the proportion of dependent people to the working population

The **Working Population** is the number of people who are at working age

Economically Active people currently work

Dependent People rely on the working population. They could be the elderly or the young

Image 6: Crowds on George Street, Sydney, Australia

Population Pyramids are used to show the age and sex structure in a country

Different Population Groups
The population consists of the working population, youth population and elderly population. The working population could be low due to less people being born, potentially leading to reduced economic growth.

Reasons for a:

	Youth Population	Elderly Population
Large	<ul><li>High Birth Rates</li><li>Low Infant Mortality</li><li>Improved Sanitation</li></ul>	<ul><li>Better Health Care</li><li>Longer Life Expectancy</li></ul>
Small	<ul><li>Lifestyle Changes – Low Birth Rates</li><li>Potential High Infant Mortality</li></ul>	<ul><li>Poor health care</li><li>Shorter Life Expectancy</li></ul>

22

1.3 POPULATION STRUCTURE

Problematic implications of:

An Aging Population	Too many Youths	Too few Youths
- Less workers to support aging population - Increased pressure on health and care services from the elderly population - Expenses from pension claims - Higher dependency ratio	- Strain on Nursing, Education and Health Services - Strain on Educational Services - Increased dependency ratio	- Potential Closure of Nursing, Education and Health Services - The population is more elderly - Potential for population decline - Less future consumers, tax payers and workers

Advantageous implications of:

An Aging Population	Too many Youths	Too few Youths
- Reduced crime – can you imagine an old lady robbing a bank? - Old people don't have to commute, so strain on transport services will be reduced	- There will be more jobs in caring and teaching - A large future workforce, more tax payers in the future and more potential consumers - More of the population would be integrated into modern technology	- There will be less Nursing / Educating/ Health Service costs - There will be a lower dependency ratio

THEME 1
POPULATION AND SETTLEMENT

Population Structures and Economic Development

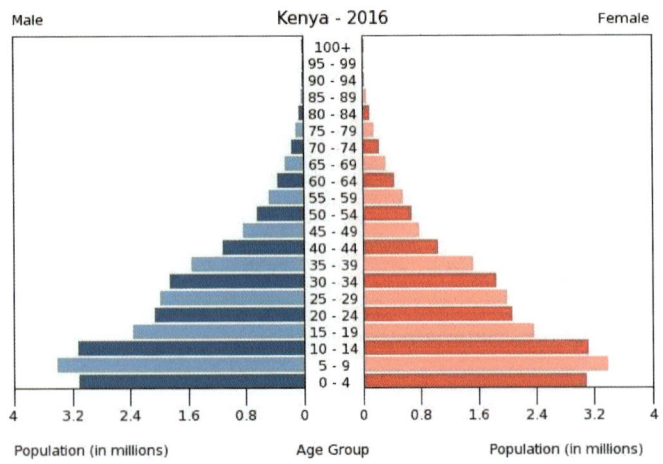

Figure 2: Kenya's Population Structure in 2016

LEDCs
Larger Youth Population

This population pyramid shows typical population structure features in LEDCs. Indicated by the pyramid's narrow top, there appears to be few elderly people - so there is a shorter life expectancy. This could be due to poor sanitation.

The wide base represents a high birth rate – as there is a large youth population. The recent reduction in births could be due to a poorer infant mortality rate or improved contraception availability.

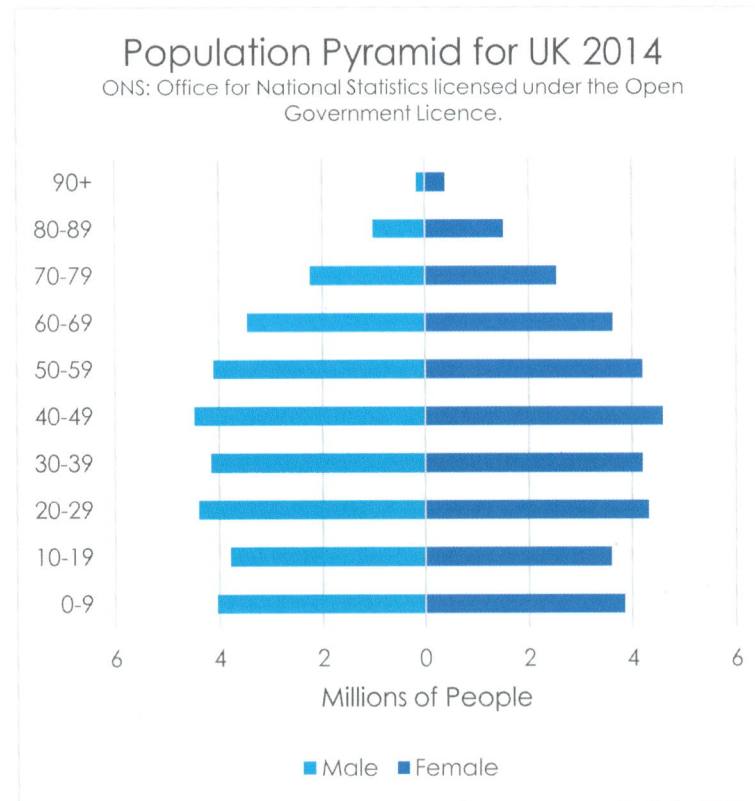

The UK's Population Structure in 2014

MEDCs
Larger Elderly Population

This population pyramid for the UK shows characteristics of an MEDC population structure. Notice that the base is narrower than in an LEDC (above), due to lower birth rates, with less youths between 10-19 years of age.

There is a wide central part, showing a large working population.

There is also a wider top part than in LEDCs, as people are living longer – creating an increasingly elderly population.

1.3 POPULATION STRUCTURE

Case Study - Japan
A country with a high dependent population

As you can see from the Population pyramid, Japan has many people above the age of 60, yet very few below 20 in comparison

There are lots of elderly dependents above the age of 60 / 70. The number of elderly people is increasing, due to long life expectancy, which causes the following problems:

- Pressure on the Japanese health services to provide medical care
- Pressure on the Japanese Government to pay pensions
- Pressure on public transport services, as much of the elderly population no longer drive

However, there is an additional problem that is potentially more important than the large dependent population. There are very few youths being born in Japan, as you can see from the tapering out of the population diagram at the base.

This could be because less women are having babies due to lifestyle changes – they may get a better standard of living without children. Or perhaps they may struggle to handle children alongside a job in one of Japan's high-tech industries -

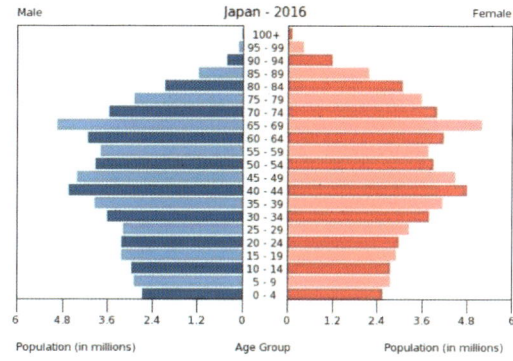

Figure 3: Japan's Population Structure in 2016

more Japanese women are involved in industry than before.

There will soon not be enough of a Japanese working population to support the elderly population.

This may lead to:

- Higher Taxes to pay for old age pensions and medical care
- A smaller population in the future
- Reduced future economic growth, as less people are of working age

The population won't be replaced, so Japan is likely to need to look at migration to provide a sufficiently sized workforce; or, Japan could attempt to increase the fertility rate by offering incentives to couples to have children. The population won't be replaced because the current fertility rate is well below the required level to prevent population decline and to maintain the population size.

THEME 1
POPULATION AND SETTLEMENT

In summary:

- Life expectancy is growing in Japan.
- So, more people are living longer
- They put more strain on healthcare and transport services
- The working population must pay for these services through taxes
- But, there are less children due to a lower fertility rate
- This is because of the increase of opportunities for women in high-tech industries in Japan
- There will be less people working, as more people retire and less take their place
- So, there will be less people to pay taxes and provide economic growth
- In the future, this will raise taxes and reduce economic growth as the population declines
- Migration could be looked at to increase the working population and to help to negate these issues
- The government could also offer incentives to couples to have children and increase the working population

Image 7: Ochanomizu, Tokyo, Japan

1.4 POPULATION DENSITY AND DISTRIBUTION

1.4 POPULATION DENSITY AND DISTRIBUTION
Describe the factors influencing the density and distribution of population

Physical, economic, social and political factors

Population Density: The number of people living in an area

Sparse Population: When there are few living in an area e.g. Sahara

Dense Population: When a lot of people live in an area e.g. London

Population Distribution: How the population is distributed in an area (even = regularly)

There are many factors that influence the population density and distribution. Dense population factors encourage people to settle, and sparse factors encourage people not to settle.

Factor	Sparse Population	Dense Population
Physical	<ul><li>Mountainous area – harder to build roads / houses on</li><li>Extreme climates – like deserts – no water / crops</li><li>Heavily Forested Area – like rainforests – hard to access</li><li>Flood Plains – houses get destroyed</li><li>Shortage of resources</li><li>Infertile land</li><li>Frequent natural disasters – e.g. earthquakes</li></ul>	<ul><li>Areas with good access – e.g. trading at coasts or rivers</li><li>Stable climate</li><li>Flat – easy to build on</li><li>Near good water supply – e.g. Thames / Nile for washing, drinking, transport or waste</li><li>Available Natural Resources</li><li>Fertile land</li><li>Not many natural disasters</li></ul>
Economic	<ul><li>No Jobs</li><li>Poor utility provision – life would be difficult</li><li>Poor communications</li><li>Poor transport systems</li></ul>	<ul><li>Lots of available jobs</li><li>Good utility provision</li><li>Good communications</li><li>Good transport systems</li></ul>
Social	<ul><li>High Crime</li></ul>	<ul><li>Low Crime</li></ul>
Political	<ul><li>Corrupt country / government</li><li>Poor public services – like education</li><li>Civil War</li></ul>	<ul><li>Generous government – building grants</li><li>Good public services</li></ul>

THEME 1
POPULATION AND SETTLEMENT

Case Studies

A densely populated country or area (at any scale from local to regional)

A sparsely populated country or area (at any scale from local to regional)

Densely Populated – Greater London Area, UK

	Factors for the Dense population distribution
Physical	<ul><li>Good Access to the region – through motorways, via the Thames, by rail or through one of London's many airports. This is partly aided by the flat terrain – meaning there are few obstacles to overcome when setting up transport links</li><li>Low frequency of extreme weather events</li><li>This part of the UK is mainly very flat, making it easier to build houses and offices</li><li>Near the Thames – a good water supply. Historically used for washing, drinking and waste. Now also used for transport. Aided early development of settlements in the Greater London Area</li><li>Natural resources available</li><li>There aren't many natural disasters in the UK</li></ul>
Economic	<ul><li>Lots of available jobs – especially in banking or trade</li><li>Good utility provision from the National Grid and the Roman sewer network</li><li>Good communication infrastructure – e.g. abundant internet access and phone masts / coverage</li><li>Excellent Public Transport systems – mostly part of TFL (transport for London)</li></ul>
Social	<ul><li>Low crime rates compared to some other MEDCs – e.g. Spain</li></ul>
Political	<ul><li>Generous British government – building grants, housing schemes, transport improvement schemes</li><li>Good public services – Education (for instance, Universities), Transport, Parks and Amenities</li></ul>

These factors have historically drawn, and currently draw, people to the Greater London region. The more people that live in a certain area, the denser the population is – so London has a Dense Population Distribution.

1.4 POPULATION DENSITY AND DISTRIBUTION

Sparsely Populated – Himalaya Mountains, Asia

	Factors for the Sparse population distribution
Physical	- The Himalaya Range is a very mountainous area, so it is hard to grow crops, build roads, build houses or easily access water
- The Himalayas have an extreme cold, icy, windy and snowy environment which makes staying warm difficult. Also, the low oxygen levels at higher altitudes make breathing harder, so building or any real activity is more difficult. This also makes it harder to grow crops – because of the cold environment at high altitude, so food availability becomes an issue. These conditions are consequently hostile to humans.
- It is also hard to access the Himalayas due to their high altitude and dramatic terrain. This prevents road building, which in turn makes it very difficult to build settlements as materials can't be transported. Also, it will be harder to get food and other resources needed for living to the area
- There will be resource shortages due to access issues
- The land is infertile, as it is mainly snow, ice and rock
- There are frequent natural disasters like blizzard storms, avalanches and Earthquakes
Economic	- No jobs, as nobody lives here
- Poor utility provision, as it is near impossible to implement utility systems in such a mountainous area
- Poor communications due to the remote and isolated nature of the Himalayas
- Poor transport systems because roads can't be built. The only reasonable mode of transport is by foot or helicopter
Social	- Nobody lives here!
Political	- No public services

A combination of these factors means that the Himalaya Mountains are uninhabitable, so the population is very Sparsely Distributed.

THEME 1
POPULATION AND SETTLEMENT

THEME 1: POPULATION AND SETTLEMENT

SETTLEMENT

1.5 Settlements and Service Provision

1.6 Urban Settlements

1.7 Urbanisation

🌍 **Case Studies**

THEME 1
POPULATION AND SETTLEMENT

1.5 SETTLEMENTS AND SERVICE PROVISION

Explain the patterns of settlement

Describe and explain the factors which may influence the sites, growth and functions of settlements

Give reasons for the hierarchy of settlements and services

A **Settlement** is a place where people live. It can be of any size – from a small isolated dwelling to a World City

The **Site** is the location which a settlement is built at. Features include altitude and resources

Situation is the position of the settlement in relation to the surrounding area. The situation's features often promote growth – like agriculture and transport

Patterns of Settlement

Explain the patterns of settlement

Dispersed, linear, and nucleated settlement patterns

AKA: **Morphology** of Settlements.

An **isolated settlement** is a single building.

A **dispersed settlement** is a settlement made up of isolated individual buildings - that are separated by a few hundred metres with no focus. It is more of an area containing buildings than a single settlement, meaning that there is a sparse population distribution. The very small settlement size means there are often few services in a dispersed settlement.

A dispersed rural settlement could form because of:

- The area being mountainous or remote
- The area mainly being used for farming – meaning there are only a few, dispersed farmhouses
- Limited building materials
- No job opportunities
- No public services

Linear settlements are settlements which form a line or arc shape. They often follow a road, valley or water body. This enables the settlement to more effectively utilise transport routes (increasing access) provided by a road or water body. They can also form due to confining steep valley sides.

1.5 SETTLEMENTS AND SERVICE PROVISION

A **nucleated settlement** is where all the buildings in the settlement are built close to each other, often around a common centre – like a junction or park. Big cities are often nucleated, meaning that they are usually well planned or structured. Reasons for nucleation could be due to:

- Good transport links – excellent accessibility
- Fertile land around the settlement, to grow food for citizens
- Flat relief means the land is easy to build on
- Lots of nearby resources
- The site being a bridging point
- The site being in a defensive or strategical location
- Good water supply
- No restrictions to development in any direction
- Good job opportunities
- Good public services

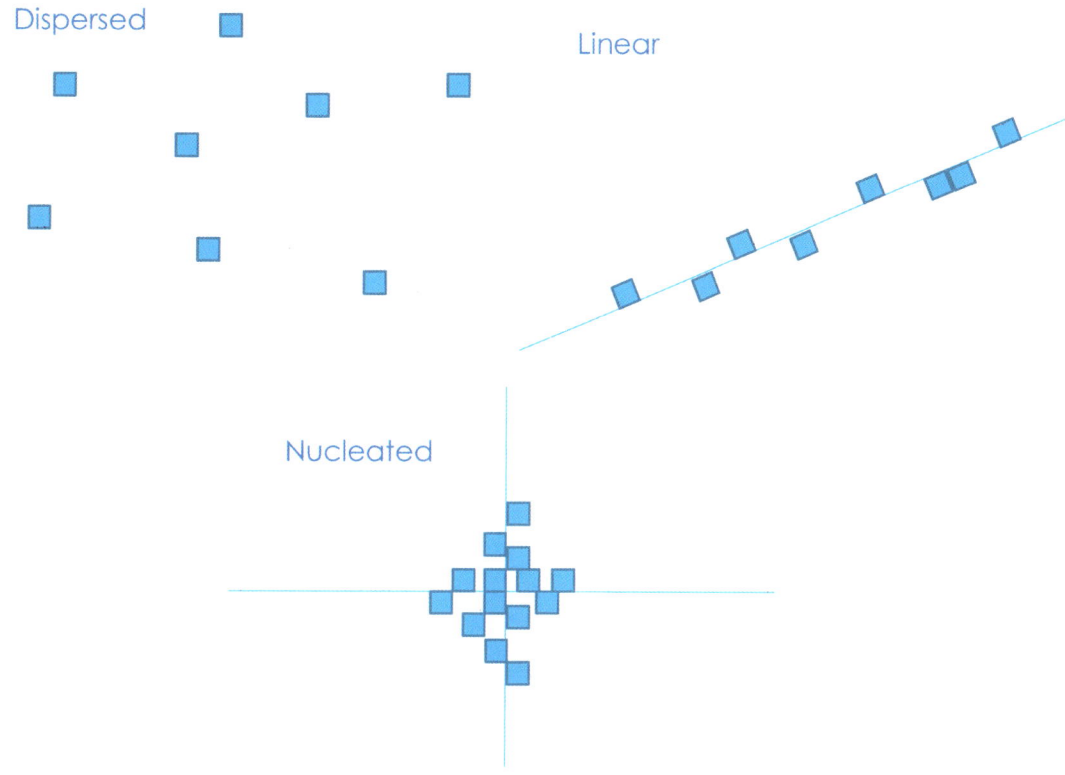

Types of Settlement

THEME 1
POPULATION AND SETTLEMENT

Factors influencing settlements

Describe and explain the factors which may influence the sites, growth and functions of settlements

Influence of physical factors (including relief, soil, water supply) and other factors (including accessibility, resources)

	Factors

Physical	- **Relief** – the shape of the land
	  - Effects how the houses are placed
	  - Flat land (flat relief) is good to build a settlement on
	- The **aspect** of the land helps to shelter crops or provide more sun
	  - Means you can yield better or more crops
	- Soil – **Fertile Soil** makes farming more productive (higher yield), meaning it would promote the formation of a farming settlement
	- **Water Supply** – provides water for living and growing crops. Is essential for any settlement

Other	- **Accessibility** / Transport – This could be increased by the presence of roads or water bodies. This could allow settlement growth by permitting importing goods, internal migration to the settlement and access to land for expansion - **Resources** – both for building houses and for potential industry. Industry promotes growth of a settlement by bringing workers into the settlement (they may live there or use the settlement's services), as well as enhancing economic growth

Factors and the settlement's function:

- **Agricultural** – good soil, aspect, accessibility and water supply – needed to grow and export crops
- **Industrial** – accessibility (to import raw materials and to transport products); market nearby (to sell products); smooth and flat relief (makes it easier to build factories); cheap land (to increase profit – saves money); existing industries (some of the materials needed for manufacturing might be produced locally); available water (to cool machinery or for production of the product) and lots of resources available (raw materials used when manufacturing products or for power generation)
- **Residential** – accessibility (for commuters) and flat relief (to cheaply, quickly and easily build housing)

1.5 SETTLEMENTS AND SERVICE PROVISION

Hierarchy of Settlements and Services

Give reasons for the hierarchy of settlements and services

High-, middle- and low-order settlements and services. Sphere of influence and threshold population

Settlement Hierarchy

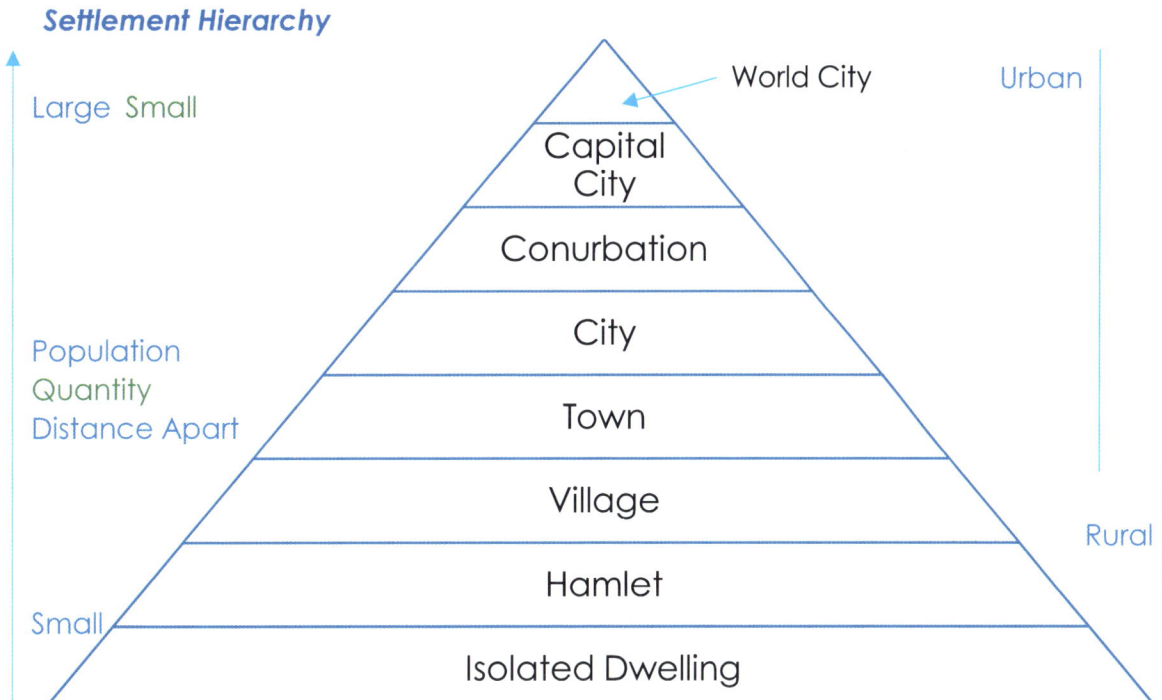

A Settlement Hierarchy is a way of putting settlements in order of importance – e.g. smallest to largest.

As you move towards the bottom of the settlement hierarchy, the number of that type of settlement in an area increases. For instance, you only have 1 capital city in a country (which is at the top of the hierarchy), yet you have hundreds of hamlets and villages.

3 key factors affecting the hierarchical position of a settlement:

1. **Population Size** – the larger the population, the higher up on the hierarchy it is; this is as there are more houses
2. The **range and number of services** – the more services a settlement has, the higher up on the hierarchy it is – as more people live there to access the services
3. The **sphere of influence** – the bigger the sphere of influence, the higher up it is – possibly because the city offers exclusive services that people may want to live near

35

THEME 1
POPULATION AND SETTLEMENT

These 3 variables are linked. For there to be a service, there needs to be a certain number of people (called the threshold population) wanting to access that service. People are more attracted to a settlement with more services. Because more people are attracted to go there (as there are more services), the sphere of influence is greater.

NB A **rural settlement** is a small settlement in the countryside.

Services
Useful Terms

Services are facilities that are offered to people – like a bakery or school. They have Threshold Populations

A **Threshold Population** is the minimum number of people needed for a service to be offered or to be available

The **Sphere of Influence** is the area that the service serves

A **Range** is the distance people are willing to travel to buy a product or service

High Order Goods are goods that people buy less frequently and are often something that is rarely bought – e.g. a sofa, phone or oven. They are often more expensive, so people will often shop around (**compare**). They will look at different products, or different prices, available at more than one service (e.g. shop).

Low Order Goods are goods that people tend to buy daily – like milk, bread and cereal. These are usually cheap. People will often buy them locally, without comparing quality or price. They are bought at services that are **convenient** (in location) for the consumer.

The number of services that a settlement provides increases with settlement size.

Small settlements (hamlets and villages) will usually only provide convenience Low Order services. Examples include a post-office, corner shop and newsagent.

Large towns, cities and other high-order settlements will provide both Low Order and High Order services – like furniture stores, electronics shops and large public libraries.

This means that larger settlements have a larger sphere of influence than smaller settlements. So, they serve a larger area.

World Cities, like London, have a global sphere of influence. Villages have a local sphere of influence.

Different services have different threshold populations. For instance, a Furniture Store selling high-order goods (in this case, sofas) will have a higher threshold population than a Corner Shop (a low order service) selling Low Order goods (like milk and sweets). This Furniture Store would need a larger number of

1.5 SETTLEMENTS AND SERVICE PROVISION

people than the Corner Shop to support it and to provide a profit. This means that the Furniture Store (a high order service) would only be found in a larger settlement, where there is the threshold population required for the service to gain a profit.

In addition, the maximum distance people would be prepared to travel (range) for the service would be very low for a Corner Shop, as it sells low-order goods. The range of a Furniture Store would be much greater, as people would be wanting to 'shop around' for the best price – as they are shopping for High Order goods.

NB A **High Order settlement** is a settlement with a large sphere of influence; and a **Low Order settlement** is a settlement with a small sphere of influence

NB **Middle Order** settlements, services and goods have a sphere of influence / value of a moderate size – in between High and Low Order. Middle Order goods are purchased from time to time. An example would be a calculator, movie or CD.

Summary

	Settlement	Service	Goods
High Order	• Has High Order Services • Provides Lots of Services • Large sphere of influence • Top of Settlement Hierarchy	• Sells High Order Goods • Large Range • High Threshold Population	• Purchased Rarely • Expensive • People compare prices / quality • Larger Range • TV, Sofa or Fridge
Middle Order	• Has Middle Order Services • Provides Some Services • Medium Sphere of Influence • Middle of Settlement Hierarchy	• Sells Middle Order Goods • Medium Range • Medium Threshold Population	• Purchased Now and Again • Pricey – but relatively cheap • People go a little distance • Middle Range • Calculator, Movie
Low Order	• Only Has Low Order Services • Provides Few Services • Small Sphere of Influence (Local) • Bottom of Settlement Hierarchy	• Sells Low Order Goods • Very Small Range • Low Threshold Population	• Purchased Frequently • Cheap • People buy them locally for convenience • Small Range • Bread, Milk, Eggs

THEME 1
POPULATION AND SETTLEMENT

Case Study – Isle of Man
Settlement and service provision in an area

The Isle of Man is a small island located halfway between England and Ireland

Douglas, the capital of the island, provides services for much of the Isle of Man. It is a high-order settlement, containing a high street, numerous retail parks, sport facilities, 2 cinemas and other High Order services.

Many people travel a fair distance to Douglas to purchase a Television, Phone or other High Order goods – as this is where they are available. Douglas has several High Order Services that sell High Order goods. An example is Currys and Pc World, a popular shop amongst Manx consumers for purchasing technology and household appliances. People travel from all over the island to shop here, as the High Order goods aren't available as cheaply, or with as much variety anywhere else on the island. They can compare goods by looking for products at alternative retailers – including Manx Telecom and Tesco. This means that these High Order services – some available in the Spring Valley Industrial Estate or Eden Park (mixed-use commercial, business and industrial parks located in the suburbs of Douglas) – have large ranges and spheres of influences.

Middle Order goods are available in Middle Order services across the island. These are often meals out (for instance, the wide selection of restaurants in Castletown), or perhaps cosmetics (available in many pharmacists in Peel, Castletown, Port Erin and Ramsey – all Middle Order Settlements). There are more Middle Order Settlements – like Castletown – than High Order settlements (there is only 1, Douglas). So Middle Order goods and services have smaller ranges than High Order goods and services.

Low Order goods are readily available across the island in Low-Order settlements (especially in hamlets, villages and towns). People don't like to travel far to get basic, low value goods – and are unlikely to want to compare them. They travel to their local Convenience Store (e.g. a Spar or Co-Op) or Post Office (Low Order services) to stock up on milk, tea, eggs and bread (all Low Order Products). There are lots of Low Order Settlements on the Isle of Man, much more than higher order settlements – which follows the Settlement Hierarchy Model.

1.5 SETTLEMENTS AND SERVICE PROVISION

Case Study – Greater Manchester, United Kingdom

The Greater Manchester area has many services. In the area, there are 3 main High Order services:

- **Manchester Airport** – A commercial and transport service
    - Its sphere of influence covers most of North West England, providing international air travel to the area. People travel a fair distance to Manchester Airport for special flight routes (e.g. to Beijing) or for the best flight deal – meaning it has a large range. Manchester Airport has a large threshold population, as lots of people need to use the airport to create a demand for flights. Its location is outside of the main city centre; also, it is near major roads; increasing accessibility.
- **Manchester University** – An educational service
    - Undergraduates from all over the country, and even the world choose to study at Manchester University because the wide range of courses available. Manchester University does have some specialised courses - along with world-class educational facilities – meaning that its sphere of influence is national, if not global. The threshold population is also high, as students are needed to fund the university. The location is in the centre of Manchester, to provide good access to the railway network, and as it was historically located there.
- **The Trafford Centre** – A commercial service
    - People travel from all over the North of the UK to buy High Order goods at the Trafford Centre (a large shopping centre). These goods include high quality clothing, TVs and watches. It is in the suburban area of the Greater Manchester region, providing better access to the shops (as it is closer to motorways). There are numerous shops here that offer similar products, allowing the consumers to 'shop around' and compare products for both quality and price.

These are just a few examples of High Order Services in Greater Manchester, meaning that it is a High Order settlement.

Low Order and Mid Order goods and services are available locally in each of Greater Manchester's districts. They are many more than the few High Order Services in Manchester – this is as they have low threshold populations, ranges and spheres of influences

Image 8: Aircraft at Manchester Airport Terminal 2

39

THEME 1
POPULATION AND SETTLEMENT

1.6 URBAN SETTLEMENTS

Describe and give reasons for the characteristics of, and changes in, land use in urban areas

Explain the problems of urban areas, their causes and possible solutions

Land Use

Describe and give reasons for the characteristics of, and changes in, land use in urban areas

Land use zones including the Central Business District (CBD), residential areas, industrial areas and the rural-urban fringe of urban areas in countries at different levels of economic development

The effect of change in land use and rapid urban growth in an urban area including the effects of urban sprawl

Characteristics of Land Use
Land Use Zones:

The **Central Business District (CBD)** – The centre of the city, usually dominated by high rise, high density buildings that contain offices, banks and other commercial or business services.

> Government buildings; company headquarters; theatres; hotels; restaurants; public transport; few residents; high land value; high order shops; congestion charge; tightly-packed streets

The **Transition Zone** – Where Factories / Industry are located

The **Inner City** – Historic Section. Some older industry and lower class residential with a high density.

> Industry; high density housing (terraces) of poorer quality, built in the population boom for factory workers; pollution – industry is located here because the inner city used to be at the edge of town (before rapid urban development outside of the inner city), it is a poor location for industry today. Often, there are more immigrants

The **Inner Suburbs** – Middle Class Residential, Medium Density

> Semi-detached houses built to house the growing population. Some houses have driveways / garages

The **Outer Suburbs** – Upper Class Residential, Low Density

> More modern, more detached houses with gardens and driveways, built on estates. There are also some small shopping centres;

1.6 URBAN SETTLEMENTS

more house owners – with the houses also being low density; few services and large open spaces

The **Rural-Urban Fringe** is between the edge of the outer suburbs and the countryside

Mixed farming / housing, Shanty Settlements in LEDCs

Land use models are theories that try to describe and explain the structure of urban city areas. Most models suggest that land value is highest in the centre of a city (as space is at a premium) – explaining why there are mainly high-rise buildings in the CBD. The low-density buildings are consequently on the edge of the city (the suburbs), where there is more space available.

One popular model, that applies to MEDCs, is the Burgess Concentric Zone Model.

LEDCs tend to have a city structure with a 'wedge' of upper class residential, 'wedges' of industry following main roads and 'informal settlements' (*also known as slums, favelas or squatter settlements*) surrounding the city – usually situated at the Rural-Urban fringe.

Burgess' Concentric Zone Model (MEDCs)

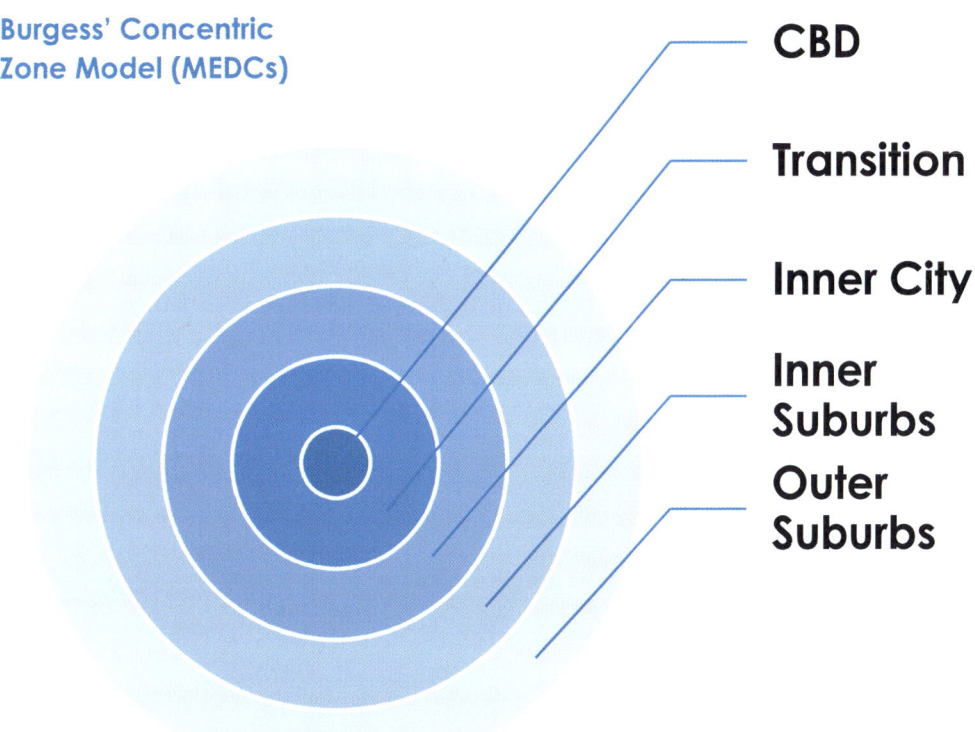

- CBD
- Transition
- Inner City
- Inner Suburbs
- Outer Suburbs

Burgess' Concentric Zone Model

41

THEME 1
POPULATION AND SETTLEMENT

Types of Land Use

Commercial - land used for business – frequently offices (this is usually in the CBD).

Residential - land used for housing – apartments are typically found near the CBD and bigger residential houses in the Suburbs.

Industrial - land used for factories – often in the transition or inner-city zones. They are now more frequently located on the Rural-Urban fringe.

Agricultural - land used for farming – and occurs outside (or on the boundary of) cities.

Retail - land used for shops. More land used to be used for retail in the CBD, but recently shopping centres are locating on the Rural-Urban fringe due to the cheaper sites and increased accessibility.

NB When asked to identify land use in the exam, say the actual purpose it is used for – e.g. park or office

Changes in Land Use

Greenfield Land – land that was previously farmland or was undeveloped

Many commercial buildings and factories are being moved outside of the main town to provide better access for both workers and customers (facilitating increased trade). This means that more land in the city centre is becoming disused (brownfield land). Also, greenfield land is being destroyed in the locations these services are moving to. The process of moving out of the CBD or city is called relocation to the Rural-Urban fringe. The Rural-Urban fringe is where most modern retail or industrial parks are now located.

In most urban areas, the land use changes most at the Rural-Urban fringe. The urban sprawl causes large scale development of roads, retail parks and housing on greenfield land, as explained above. Developers decide to build here because land is more accessible (there are nearby motorways), abundant and cheap than inside a large urban area.

More and more residential areas are also moving to the suburbs, to provide residents with larger houses and more living space – creating a more laid-back suburban feel.

Effect of rapid urban growth

Rapid urban growth accelerates the construction of commercial and retail areas on the Rural-Urban fringe, destroying greenfield land.

1.6 URBAN SETTLEMENTS

The relocation of industry and commercial services from the city centre to the rural-urban fringe poses the following issues:

- The CBD becomes 'dead' – AKA 'the doughnut effect' – as all the services start to surround the city rather than be inside the CBD
- Disused brownfield land is created in the city centre
    - This land stays undeveloped, as it is easier and cheaper to develop on greenfield sites
    - So, the city centre becomes more derelict and neglected – potentially becoming a crime hotspot
- Greenfield land is destroyed when people build new retail and industry parks
- Countryside lost as the city expands into the countryside
- Traffic increases in the area, creating more pollution and congestion

Problems of Urban Areas

Explain the problems of urban areas, their causes and possible solutions

Different types of pollution (air, noise, water, visual), inequality, housing issues, traffic congestion and conflicts over land use change

Image 9: Smog in Cairo

As cities increase in size and population, they become more crowded, stressful and difficult to live in. Previously pleasant neighbourhoods may become **undesirable** to live in due to high crime rates. Many cities weren't planned when initially constructed to support the large populations they support today, meaning that the city struggles to support the large number of residents. So, it becomes harder to provide the services and accommodation for these people, along with jobs, that are needed to support rapid population growth.

43

THEME 1
POPULATION AND SETTLEMENT

1. Pollution

- Air Pollution
    - Traffic – releases carbon dioxide, leading to severe air pollution
    - Factories release pollutants like sulphur dioxide and nitrogen oxide, creating acid rain and haze above the city
    - Power stations burning fossil fuels release carbon dioxide

    SOLUTIONS: Introducing a **Congestion Charge** to reduce traffic; putting forward new legislation to prevent industry from being in city centres; enforcing the allowed amounts of pollutants that can be released

- Nosie Pollution
    - Traffic is noisy
    - Industry and power generation also makes lots of noise

    SOLUTIONS: Introducing a Congestion Charge to reduce traffic; putting forward new legislation to prevent industry from being in city centres; enforcing operational hours of heavy industry, to limit times of heavier noise pollution

- Water Pollution
    - Rivers used as a sewerage system may carry untreated sewage
    - Industry may dump waste chemicals in rivers
    - Toxic chemicals may be disposed of in rivers

    SOLUTIONS: Investing in improved water supplies; creating more rigorous sanitation policies and strategies; enforcing the allowed amounts of pollutants that can be released

- Visual Pollution
    - From unsightly buildings or industry in city centres

    SOLUTIONS: New legislation to prevent industry from being in city centres; more rigorous planning permission criteria

2. Inequality

Rich people get better value land than poor people, and get to commute over a shorter distance. This may be because good land (with good accessibility) is more expensive, so only the rich can afford it. This could limit the job opportunities of poorer people, because they end up with substantially longer commutes than richer people if they want to work in some better jobs.

SOLUTION: Building of affordable housing in city centres

1.6 URBAN SETTLEMENTS

3. Housing Issues

Many houses are built quickly without any planning, perhaps illegally (quite common in some LEDCs), creating informal settlements on the Rural-Urban fringe.

SOLUTION: Government backed efforts to house the growing population in legal and well-built houses with good service access

4. Traffic Congestion

More people live and work in a city, so more commute – creating traffic on roads.

SOLUTION: Improvements to public transport (e.g. metro or tram), so the roads are cleared of commuters (they could travel by public transport instead)

5. Urban Sprawl and land use changes

CBDs and inner cities are becoming disused and derelict, meaning citizens begin to become concerned over crime

SOLUTION: Regeneration of City Centres

Urban Decay
This is when parts of a city become run-down and undesirable to live in.

For instance:

- Empty buildings being vandalised
- The demolishing of buildings creating derelict (unused) land
- Buildings falling into disrepair
- The formation of crowded and undesirable Informal Settlements with no services / sanitation

Schemes are introduced by the city council to reduce these issues. The process of improving an urban environment is called **Urban Regeneration**.

Ways to do this include:

- Retrofitting the houses to give them better facilities
- Cleaning up the area
- Landscaping
- Building new services
- Demolishing and Rebuilding poorly constructed buildings
- Encouraging migration into city centres
- Encouraging businesses to locate into the CBD or Inner-City

THEME 1
POPULATION AND SETTLEMENT

Case Study – Manchester
An urban area or areas

Land Use

Industrial: Located at the Rural-Urban Fringe – Manchester is an industrial city

Residential: Located in the Suburbs

Commercial: At the Rural-Urban Fringe and in the CBD – e.g. the Trafford Centre and the Arndale Centre

Changing Land Use

Most of the retail parks (like the Trafford Centre) are located on the outskirts of Manchester – away from the CBD – for better accessibility. This is also because the land there is cheaper and more readily available. There are lots of major roads near the Trafford Centre, meaning that people from all over the North of the UK can easily drive there. They did this so, regionally, people can compare High Order goods at the services located there, driving up profits.

However, this means that derelict brownfield sites are being created in the Inner City and CBD. This is a result of commercial services moving to the Rural-Urban fringe, while building on cheap greenfield land.

Problems

Pollution has been partially deferred by the movement of industry towards Manchester's Rural-Urban fringe. Also, the reduction of congestion and traffic (mentioned below) has meant that less car fumes and car noises are present in the city centre – reducing noise and air pollution.

Inequality has been dealt with by building affordable housing in some areas of the Greater Manchester region, accommodating people who may not be able to afford the high prices of housing in the city centre.

There used to be greater congestion in the Greater Manchester area until recently. The reason for the reduction was expansion of the Metrolink Tram service network, meaning less people needed to commute by car. Lower public transport prices encouraged people further to drop the car and hop on mass transit – achieving the council's target of reducing congestion in the city centre's narrow streets.

The movement of services to the rural-urban fringe has been reduced by regeneration projects, which aim to encourage businesses back into the CBD and Inner City.

1.6 URBAN SETTLEMENTS

Regeneration

The centre of Manchester has undergone, and is undergoing, **urban regeneration** and **urban redevelopment** to attract commercial services back into the CBD to prevent sites from becoming derelict and disused. They have done this by performing urban landscaping, making the centre of Manchester a more pleasant area for consumers and citizens.

One notable area of improvement is in the Manchester Printworks. These were redeveloped in 2000 to reduce the amount of disused spaces and to provide new commercial services in the city centre– like a cinema and restaurants.

The effect is that the urban centre is now more attractive for new businesses and citizens, regaining its commercial status.

Image 10: The Printworks, Manchester

47

THEME 1
POPULATION AND SETTLEMENT

1.7 URBANISATION

Identify and suggest reasons for rapid urban growth

Describe the impacts of urban growth on both rural and urban areas, along with possible solutions to reduce the negative impacts

What is Urbanisation?

Urbanisation is the increase in the proportion of people living in towns or cities in an area.

Rapid Urban Growth

Identify and suggest reasons for rapid urban growth

Reference should be made to physical, economic and social factors which result in rural depopulation and the movement of people to major cities

Rural Depopulation is the decreasing population of rural areas as people move to urban areas.

Urban Growth is the expansion of towns and cities – due to more buildings being built. The urban area covers more land and can support a larger population.

	Reasons for Urbanisation
Physical	- **Drought / Lack of resources** in rural areas may mean that people move to urban areas to access water services or resources for a better quality of life - **Natural Disasters** in rural areas may damage settlements, forcing people to move to urban areas. The disaster may cause them to lose their rural lifestyle and self-sufficiency, meaning they need to rely on services – pressuring them to move to urban areas

Economic	- **Jobs**
	  o Pull: People may find better paid or more attractive job opportunities in cities. As a result, they might move to an urban area for an improved, more luxurious lifestyle
	  o Push: There may not be many job opportunities in a rural area – the salaries may be low or the working conditions poor

Social / Political	- **Education** – parents may move to an urban area for their child to attain a better quality of education, which might give the child better prospects. There also may not be many schools in rural areas

1.7 URBANISATION

- **Social** – many rural areas lack leisure or entertainment services. These could be more regularly accessed you lived in an urban environment
- **Healthcare** – people might move to cities to be able to access healthcare services for a better quality of life. These services may not be as available (or as good) in rural areas
- **Safety** – there may be high crime rates in a certain area, causing people to move into a safer urban district

Rapid urbanisation primarily occurred in MEDCs during the 19th and early 20th centuries, as people moved from the rural areas into cities (mainly for work in the expanding industries). This rate of urbanisation in MEDCs has slowed since then, and **counter-urbanisation** (moving from the city to rural areas) is beginning to occur in larger cities.

Image 11: Urban Sprawl in South Los Angeles

Since the 20th Century, most of the world's urbanisation is happening in LEDCs. This is due to population pressure and a lack resources or jobs in rural areas. In LEDCs it is a much more common for people to think that life is better in a city than it is in MEDCs.

Impacts and Solutions

Describe the impacts of urban growth on both rural and urban areas, along with possible solutions to reduce the negative impacts

The effects of urbanisation on the people and natural environment

The characteristics of squatter settlements

Strategies to reduce the negative impacts of urbanisation

Urbanisation is happening at an alarming rate in NICs and LEDCs around the world. This is because people think that they will have a better quality of life in urban areas, inducing mass rural to urban migration. However, this rapid growth puts pressure on urban infrastructure: like power, water and roads.

THEME 1
POPULATION AND SETTLEMENT

Impacts on People

The rich suffer much less from this migration than the poor. They get spacious, accessible and well-built homes. They also obtain good access to work and services.

The very poor have to live in **Squatter Settlements** (see end of section), which often have very poor living conditions. Poor people from rural areas frequently end up working in the informal or labour sectors (which have a low pay with long hours).

Impacts on Rural Areas

Because of rural depopulation:

- People of working age leave
    - Families may become separated by distance
    - Old people will most likely be left behind to care for the family farm. But, because they are potentially less able, the farm's yield (and consequently, profit) may be reduced

Impacts on Urban Areas

- Strain on Infrastructure
    - Roads and public transport – leading to congestion
    - Power, water, sewerage and other utilities
    - Healthcare and sanitation – pressure on hospitals
    - Education systems – overcrowding of schools
- Noise, air and water pollution
- Job shortages
- Housing shortages

Image 12: Traffic Congestion on the I-110, Los Angeles, CA

Environmental Impacts of Rapid Urbanisation

Problems

- Air Pollution: Traffic emissions; power plant emissions; factory emissions
- Waste: Rubbish not being collected; landfills in LEDCs
- Water Pollution: Sewage and waste being dumped into rivers
- Noise Pollution: Increased traffic; heavy industry; mining

1.7 URBANISATION

Reducing the negative impacts

- Investing in better water supplies or sanitation systems
- Laws to reduce air pollution
- Putting in place responsible waste disposal systems – e.g. recycling

LEDCs – Squatter Settlements

Ever more frequently, Rural-Urban migration in LEDCs leads to the development of squatter settlements (*also known as informal settlements, slums or favelas*) on the outskirts of the city. This land normally isn't suitable for housing. Squatter settlements can be extensive and can be home to many settlers. These people are often the poorest in an urban area – in many cases, they may have just moved into the urban area, unable to afford a good quality of housing.

Image 13: Houses in an Informal Settlement

The buildings are usually made from materials found nearby - commonly corrugated-iron, mud or wood.

Living conditions in Squatter Settlements:

- High population density
- Poor sanitation – inducing a low life expectancy
- Few services – like clean water and waste disposal
- High unemployment – so people work in the informal sector
    - The squatter settlements are often far from the CBD, so people can't travel as easily for work
- The land is often unpleasant, dangerous or undesirable – not suitable for being built on

(See next page for how authorities are improving them)

THEME 1
POPULATION AND SETTLEMENT

How authorities are improving them:

- **Low-interest loans** to support homebuilders
- **Self-help schemes** to provide the tools, materials and education for people to improve their housing conditions
    - Successful in most large LEDC cities
    - For this to happen, people will need to have land ownership – this is so they feel comfortable in investing in their own house
    - People slowly improve their houses (perhaps by upgrading walls or roofing)
    - The houses can also be enlarged for more living space
    - Authorities may provide clean water, sanitation or waste collection services
    - Over time, commercial transport operators or other services may become available in the area – giving the people living there a better quality of life
    - Over time, the residents can work together to improve their area and community
    - The result is the conversion of an unpleasant, illegal squatter settlement into a more pleasant, legal residential area
- **Site and service schemes** give people the opportunity to rent or buy a plot of land with access to services and transport
    - Successful on a small scale
    - An available area of land is found – which is divided into plots by the authorities, who also set up services in the area. Migrants can then rent this land and build their own house within set guidelines
    - As time goes on, they earn money to improve their house
    - This strategy was used in Lima, Peru
    - It was cheaper than building new flats, as the people can afford to do it themselves
    - Houses are of better quality because of the restrictive guidelines
    - Availability of utilities improves sanitation and quality of life
    - However, this strategy is harder to implement in large cities, as land is less available in accessible locations
    - Additionally, unemployed people can't afford it
- Giving people **ownership of the land** their house is built on
- **Investing in rural areas** to reduce rural-urban migration

1.7 URBANISATION

Case Study – Mumbai, India
A rapidly growing urban area in a developing country and migration to it

Dharavi is a massive slum in Mumbai's city centre, covering over 3km² of former mangrove swamp. Mumbai is a rapidly expanding city, with the population over quadrupling since 1950. Hundreds of families migrate to Mumbai every day, and about half live in Mumbai's several squatter settlements (because they can't afford to live in formal settlements).

Dharavi is situated between 2 railway lines. Most of the homes are well constructed – usually made from wood or brick. Many of them even have electricity.

Why migrate to Mumbai?

- The opportunity to work in the informal sector – providing an income
- Other members of family went to Mumbai – the desire to reunite
- The illusion that life will be better in Mumbai – this is really the potential of obtaining a better quality of life

Conditions in the Slum:

- No water with only some electricity availability
- Low incomes for the residents
- Workers work in the **informal sector**, often recycling used goods
    - No regular wage – it is an unofficial job with no contract, job security, health insurance or pension
- Serious air pollution from vehicles, factories and burning of waste materials
    - The result of this is that many people living in Dharavi develop breathing problems – like bronchitis
    - Air pollution from vehicle exhaust gasses, burning waste, industry and power generation using coal
- **Water pollution in the Mithi river**, because people have no other way to dispose of their waste
    - Industry plays a role in increasing water pollution
    - Airport dumps plane oil into the river
    - Millions of litres of sewage enter the river untreated
- Fossil fuels are burnt to provide cheap energy for growing industries and the increasing

THEME 1
POPULATION AND SETTLEMENT

population – meaning that pollutants are released (including Carbon Dioxide, which leads to enhanced global warming)
- Noise pollution from people working in Dharavi
- A flood risk created by waste dumped in the **Mithi River** blocking up drainage systems

What is being done?

Vision Mumbai, a plan devised by the Mumbai authorities, aims to improve the quality of life of residents living in Mumbai's slums – including Dharavi. By doing this, Mumbai can develop further and become an even more prosperous city. The plan aims to reduce pollution, resolve water issues and reduce the number unsightly slums.

Because Dharavi is in the centre of Mumbai, it has a high land value. This means, as part of Vision Mumbai, the slum could be demolished and the land sold. Buildings of better quality (improving the quality of life of residents) could be built – most likely in tower blocks to lower costs. This could massively reduce the number of residents in squatter settlements like Dharavi.

To attract developers, the Mumbai authorities are suggesting that the land is sold at a cut-price. The land can then, after building replacement housing for Dharavi's residents, be used to build shops, offices and apartments – which can be sold at a premium to richer migrants moving into Mumbai.

However, many residents in Dharavi are worried about how their lives – their jobs and culture – are going to be affected by both the relocation and the transition from a slum style settlement into vertical housing.

As this is a widely shared concern (Dharavi is a very large settlement), the authorities need to deal with the issues raised. The pressure is building due to rapid migration, as there is nowhere for the poorer migrants to go other than the slums.

Other measures:

- To reduce air pollution, the Mumbai authorities are encouraging the use of public transport; other strategies include banning diesel (which when burnt, releases carbon particulates), and reducing traffic by streamlining road infrastructure
- To resolve water pollution, the Mumbai authorities are looking at reducing the environmental impact of slum workers and nearby industry; along with educating residents about the importance of keeping the **Mithi River** clean
- To help to reduce flood risks, dredging has been undertaken to remove waste

1.7 URBANISATION

from the Mithi river. However, this hasn't improved the water quality that much – as waste is continuing to be dumped there.

Image 14: Dharavi (near Mahim Junction), Mumbai, India

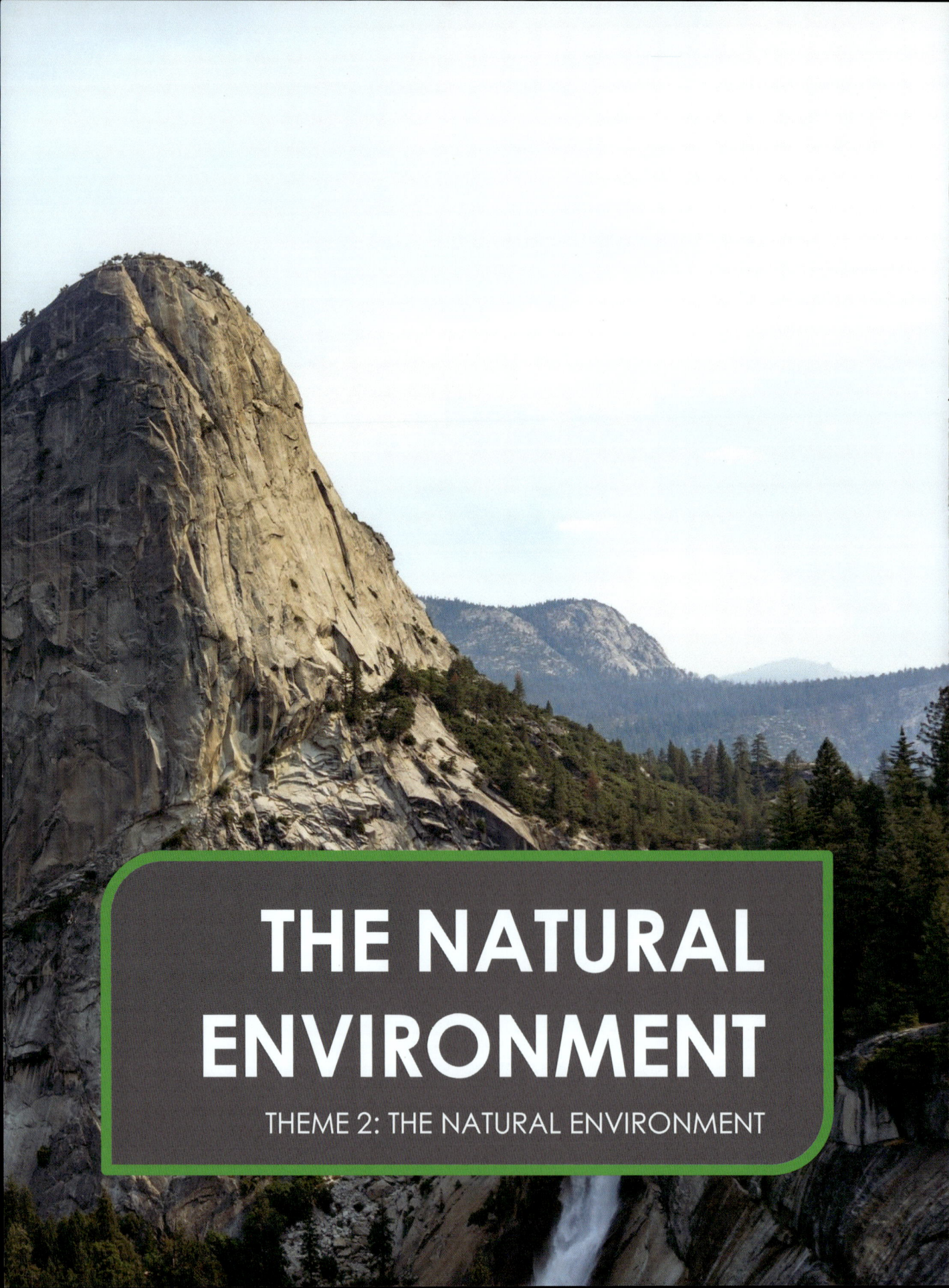

THE NATURAL ENVIRONMENT

THEME 2: THE NATURAL ENVIRONMENT

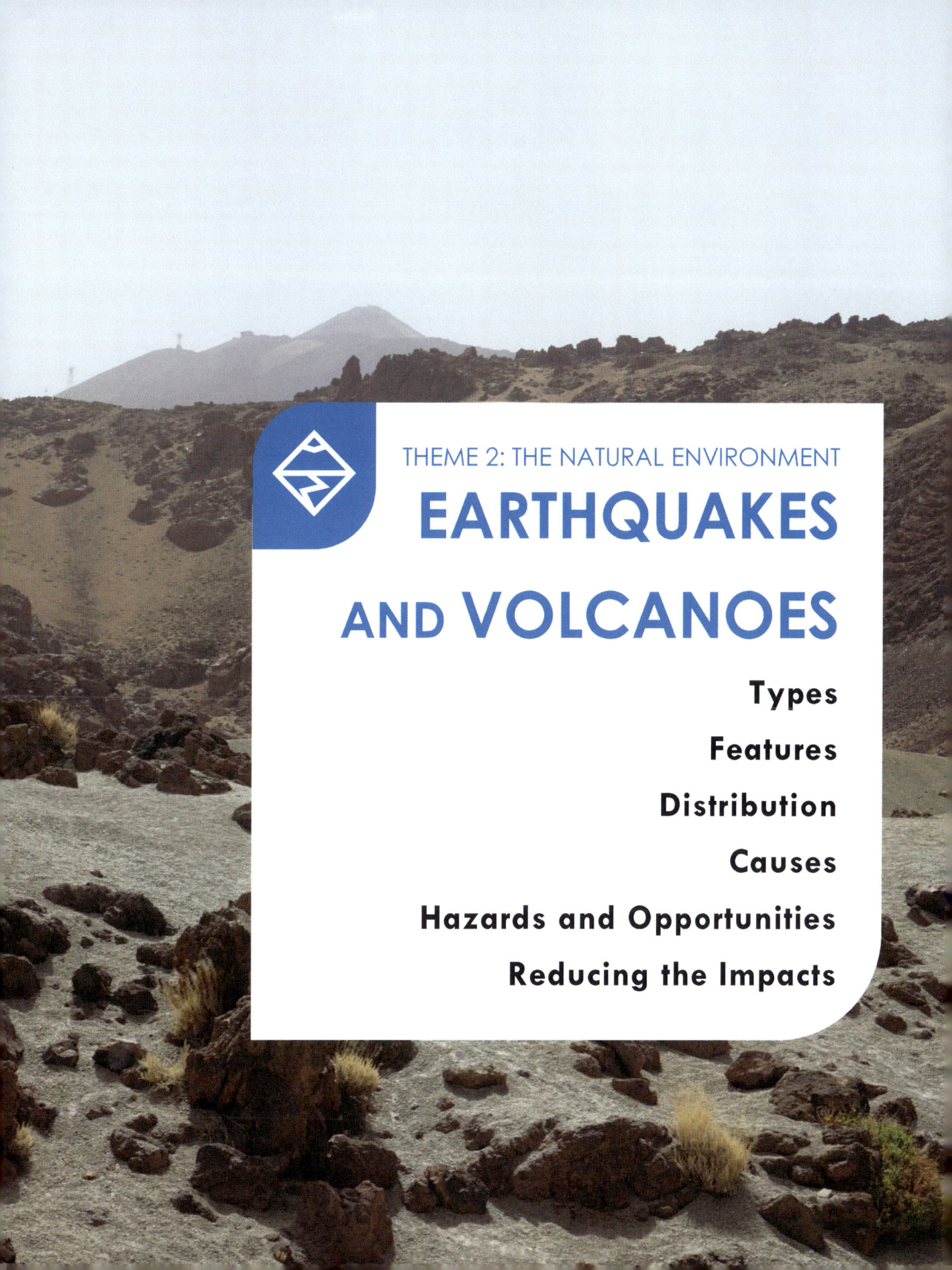

THEME 2: THE NATURAL ENVIRONMENT
EARTHQUAKES AND VOLCANOES

Types

Features

Distribution

Causes

Hazards and Opportunities

Reducing the Impacts

THEME 2
THE NATURAL ENVIRONMENT

Describe the main types and features of volcanoes and earthquakes

Describe and explain the distribution of earthquakes and volcanoes

Describe the causes of earthquakes and volcanic eruptions and their effects on people and the environment

Demonstrate an understanding that volcanoes present hazards and offer opportunities for people

Explain what can be done to reduce the impacts of earthquakes and volcanoes

Tectonics

The global pattern of plates, their structure, an awareness of plate movements, subduction zones and their effects – constructive/divergent, destructive/convergent and conservative plate boundaries

For millions of years, forces and processes have been taking place at and beneath the Earth's crust, making the Earth's surface look like it does today. We discovered and learnt more about these fundamental processes from studying volcanoes and earthquakes. Photography and satellites have also allowed us to map the ocean floor to better understand tectonic processes.

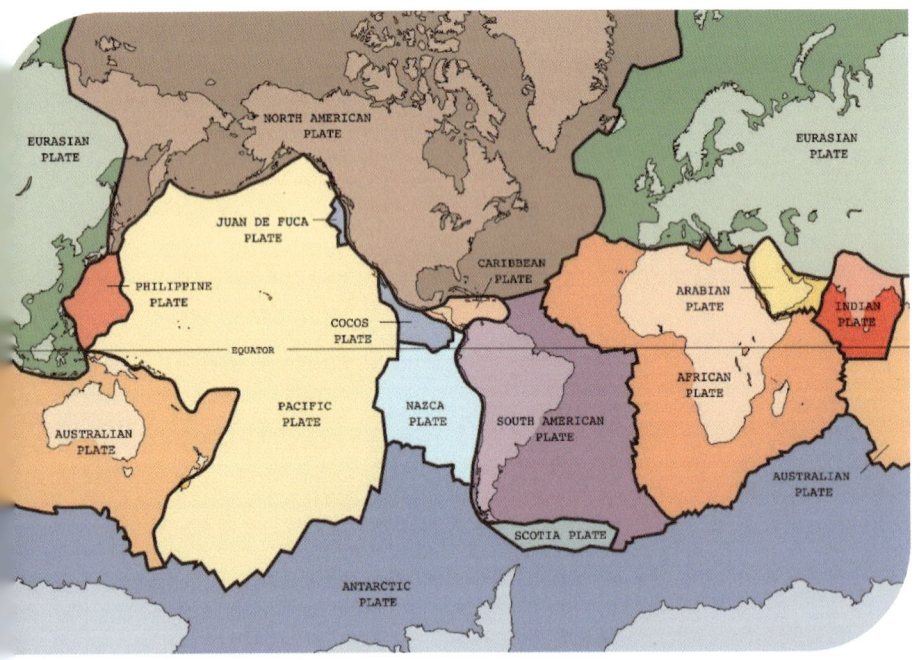

The surface of the Earth is called the crust. It is made up of massive, interlocking **tectonic plates**. The plates are moved by **convection currents** flowing in the Mantle. These currents produce all the plate movements, forming fold mountains, volcanoes and earthquakes.

Figure 4: Position of Tectonic plates around the world

2.1 EARTHQUAKES AND VOLCANOES

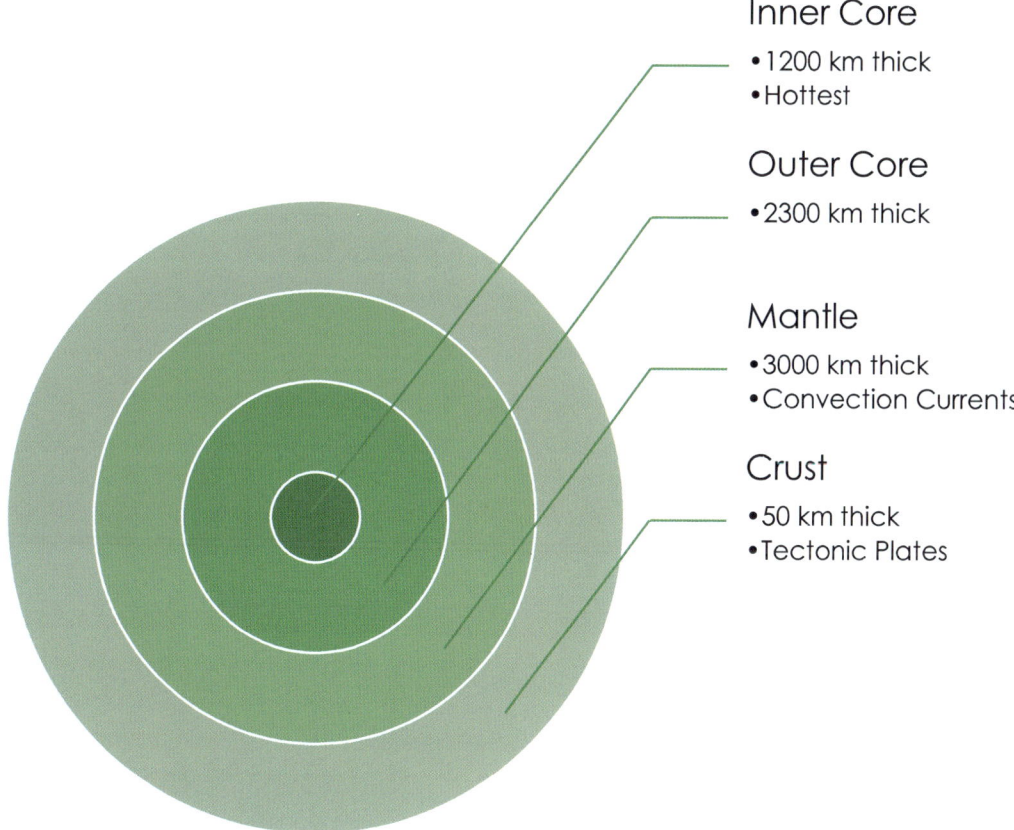

Structure of the Earth

Volcanoes and Earthquakes usually occur on plate boundaries and don't normally happen on stable tectonic plates.

Although most plates are quite stable (when the area is far from a plate boundary), they can have geological weaknesses that can create faults.

Some parts of the mantle are hotter than others and are referred to as **hot spots**. These can melt some of the lower crust, forming a weakness. Over time, the weaknesses in these spots can be worn through, meaning magma can rise to the surface, forming volcanoes. One example is in Hawaii.

Continental crust is dense and holds most of the Earth's land.

Oceanic crust is less dense and holds most of the Earth's oceans.

Plate Boundaries are where 2 plates meet.

There are 3 main types of plate boundaries: **constructive** *(divergent)*, **destructive** *(convergent)* and **conservative** *(transformative)*.

THEME 2
THE NATURAL ENVIRONMENT

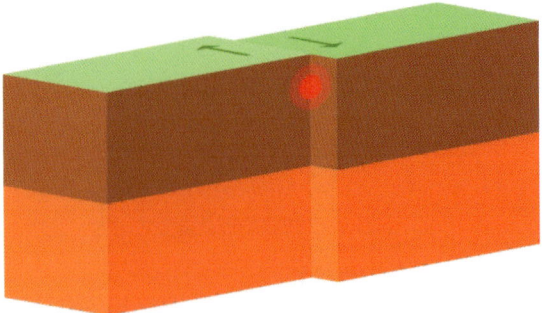

Figure 5: Conservative Plate Boundary

Conservative plate boundaries are where 2 plates move alongside each other. Earthquakes happen here due to friction between the 2 tectonic plates. An example is the San Andreas fault in California. Land isn't formed or destroyed.

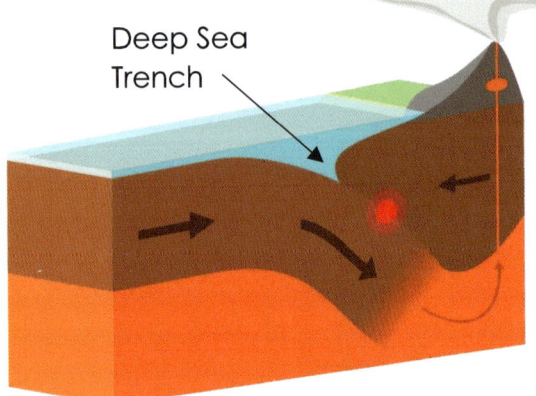

Figure 6: Destructive Plate Boundary

Destructive plate boundaries (**convergent**) are where less dense oceanic crust is **subducted** under denser continental crust. Earthquakes and Stratovolcanoes are experienced and formed here. Land is destroyed at the **subduction zone**, inducing intense heat and pressure – which forces up magma.

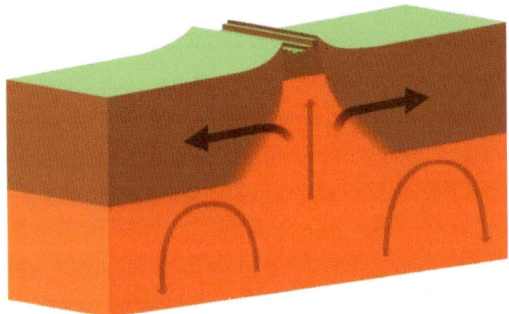

Figure 7: Constructive plate boundary

2.1 EARTHQUAKES AND VOLCANOES

Constructive plate boundaries (divergent) are where 2 plates move apart. Magma rises in the gap, cooling to create new crust. An example is the Mid Atlantic Ridge. Gentle earthquakes and shield volcanoes occur here.

Earthquakes mainly occur on constructive and destructive plate margins due to friction between the 2 tectonic plates. This friction creates vibrations through the rock, which is called an **earthquake**.

On the 3D diagrams, the red dot is the epicentre of the earthquake, where the friction occurs, and the earthquake originates from.

Volcanoes mainly occur on destructive and constructive plate margins. The rising magma sometimes escapes through the gap between the 2 plates on constructive plate margins, a **volcano** forming when it supersedes the crust. They also occur on destructive plate margins when the oceanic crust is broken down into magma at a **subduction zone**, potentially permeating (through pressure) to the surface, forming a volcano.

Continental Crust	Oceanic Crust
Older	Younger
Lighter / Less Dense	Heavier / More Dense
Can't sink	Can Sink
'Permanent'	Constantly being destroyed and formed
Contains most of Earth's landmasses	Contains most of Earth's oceans

There are also **Collision Plate Boundaries**. Here, 2 continental plates collide, forcing land upwards into fold mountains – such as the Himalayas. As neither are dense enough to sink, they are subsequently forced up. There is earthquake activity here, but no volcanic activity – as no magma is being forced up.

Volcanoes

An **erupting** volcano is a volcano that is currently releasing lava or ash in a volcanic eruption.

An **active** volcano is a volcano that has recently erupted.

A **dormant** volcano is a volcano that hasn't recently erupted but may erupt again in the future – as activity is still detected at the site.

An **extinct** volcano is a volcano that is unlikely to erupt again as no magma is rising through it any more. This is usually after about 10,000 years of inactivity.

Types and Causes of Volcanoes

Types of volcanoes (including strato-volcanoes [composite cone] and shield volcano)

Describe the causes of . . . volcanic eruptions . . .

Viscosity is how thick something is.

LOW SILICA CONTENT = LOW VISCOSITY (RUNNIER)

Volcanic Eruptions occur when tectonic activity causes a pressure build-up. This pressure is then relieved by the releasing of magma through the vent of a volcano – creating a volcanic eruption.

Image 15: Arenal Volcano

2.1 EARTHQUAKES AND VOLCANOES

Stratovolcanoes (Composite Cone Volcanoes)

Volcanoes are very explosive at destructive plate margins, releasing enormous amounts of material. The volcanoes here are made up of layers of ash and lava, which build up over several eruptions.

On destructive plate boundaries, oceanic crust is melted. The melting oceanic plate induces intense heat and pressure, forcing magma upwards. This magma is forced up through the boundaries' fold mountains, creating **Stratovolcanoes**.

Liquid molten rock (called **magma**) builds up underneath the crust in a magma chamber. Pressure from the magma against rocks in the ground causes an eruption through one of the volcano's vents. This magma then flows out (we call it **lava** above the surface). Also, ash clouds settle on the lava flows while the lava cools and sets.

Stratovolcanoes have steep sides because the viscous (also referred to as **acid**) lava cools more quickly (and flows more slowly) due to the greater silica content.

Rocks thrown out of the volcano as part of the eruption are called **volcanic bombs**. When magma leaks out through the alternating strata of lava and ash, a **secondary cone** is formed.

Shield Volcanoes

Shield volcanoes are generally much flatter than stratovolcanoes, less violent and have less dramatic eruptions. There is little pressure build-up at conservative plate boundaries, as the pressure is released in gaps created by the plates' diverging motion. Because the pressure doesn't build up as much as beneath stratovolcanoes, shield volcanoes erupt less frequently and less explosively.

When the magma is more liquid, it flows more gently and more slowly to the surface, forming a shield volcano. The eruptions are less explosive as there is a lower silica content in the lava – meaning the lava is more **basic**, reducing viscosity and increasing flow distance. This basic lava flows to the surface through cracks in the new crust at constructive plate margins.

Also, shield volcanoes can form at a 'hot spot' – meaning they are sometimes located far from plate boundaries (like in Hawaii).

The cone is low and wide, with a gentle slope. This is because the lava flows more freely (basic lava has a low viscosity), so spreads thinly over a large area.

Comparison

Type	Stratovolcano	Shield Volcano
Where?	Destructive Plate Boundaries	Constructive Plate Boundaries and Hot Spots
Formation	<ul><li>Powerful eruptions occur on a destructive plate boundary. These eruptions cause lots of material to spew out, building up on the slopes.</li><li>Material is viscous and sticks to the sides. It builds up over time in layers (alternate Lava and Pyroclastic eruptions).</li></ul>	<ul><li>Formed by runny lava that flows easily down the slope away from the summit. It has a low silica content, allowing the lava to flow more quickly. This means it has a wide base.</li><li>Often found on the deep ocean floor (where the oceanic plate is formed).</li></ul>
Characteristics	<ul><li>Steep Slopes</li><li>Tall and narrow</li><li>Very viscous lava</li><li>Alternating Lava and Pyroclastic eruptions</li><li>More Explosive</li><li>Has dormant periods</li></ul>	<ul><li>Gentle upper slopes, steep lower slopes</li><li>Short and wide</li><li>Low viscosity</li><li>Usually circular or oval</li><li>Composed of thin lava flows</li></ul>
Lava Type	Acid (High Silica)	Basic (Low Silica)
Examples	Etna, Italy	Hawaiian Islands

Features of Volcanoes

Features of volcanoes (including crater, vent, magma chamber)

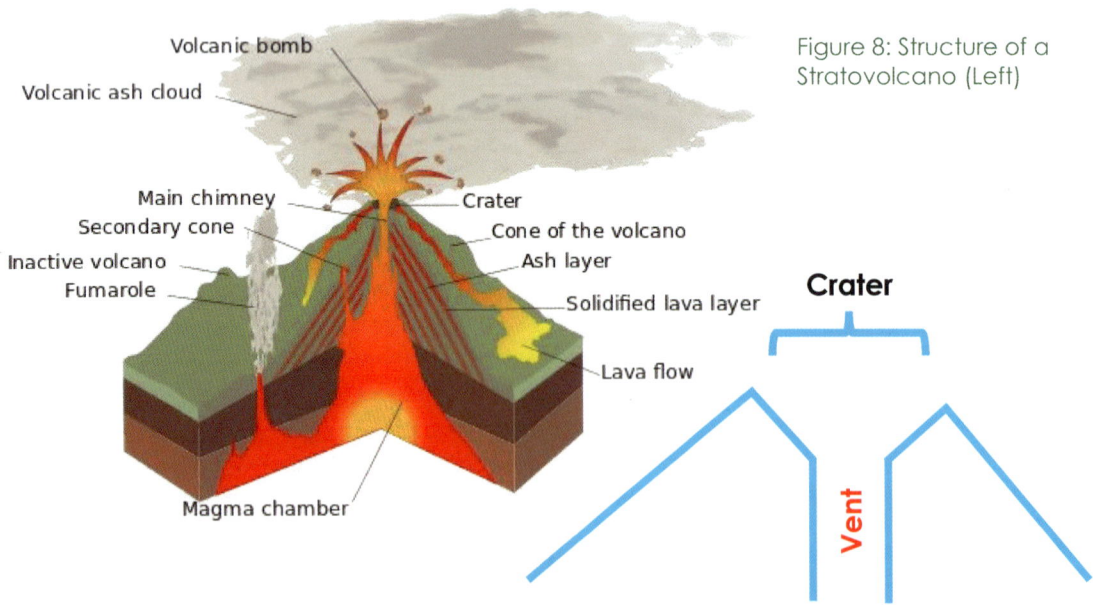

Figure 8: Structure of a Stratovolcano (Left)

2.1 EARTHQUAKES AND VOLCANOES

NB The **secondary cone**
has a secondary vent which material exits through

NB The **Main Chimney**
is another name for the (Main) Vent

NB The **Crater**
is the area at the top of the vent, a slight dip in the volcano's contour

Effects of Volcanic Eruptions
Describe . . . their effects on people and the environment

	Positive	Negative
People	<ul><li>Tourism (e.g. Iceland)</li><li>Increased farming yield due to minerals in ash</li><li>Potential for geothermal power</li><li>Can be studied to provide information about the Earth's geology</li></ul>	<ul><li>Can kill people</li><li>Can destroy buildings and infrastructure from flash flooding because of ice meeting lava</li><li>Ash clouds can devastate crops</li><li>Ash causes airspace disruption</li></ul>
Environment	<ul><li>The soil can gain nutrients from a small amount of ash (like in Iceland)</li></ul>	<ul><li>Heavy ash clouds can kill wildlife</li><li>Some volcanic gasses can cause pollution (like carbon dioxide)</li></ul>

Image 16: Mount St. Helens Erupting, USA

THEME 2
THE NATURAL ENVIRONMENT

Hazards and Opportunities
Demonstrate an understanding that volcanoes present hazards and offer opportunities for people

<u>*Why are more people becoming exposed to the hazards created by volcanoes?*</u>

More and more people are living near potentially dangerous volcanoes because of urbanisation in the 'Pacific Ring of Fire' region (regions mainly effected by this urbanisation near the ring of fire include East Asia, Central America and South America). The main hazard for people in these areas is the risk to life, property and infrastructure from ash and lava flows. In South America, volcanoes threaten people living in flatter areas. One secondary threat here is *flooding* from Andes glacial ice melting during a volcanic eruption – displacing or harming people.

Although there are hazards, volcanoes also can provide opportunities and benefits to the local people. Volcanic *soils are fertile*, thanks to minerals in the ash released in eruptions. Heat from volcanic activity also means that these areas have potential for using *geothermal energy*. Volcanoes are also a *tourist attraction* – so people are attracted to go on holiday to see them.

Hazards	Opportunities
• **Volcanic Bombs** – ejected rocks – can kill people as they fall from the sky • **Ash clouds** – cause havoc for aircraft (e.g. Eyjafjallajökull 2010) • **Lava flows** can destroy settlements • **Melting of glacial ice** – flooding, lahars and mudflows can travel long distances, engulfing settlements and destroying infrastructure	• **Fertile soils** – gives agriculture higher yields and profit • **Tourism** (referred to as *Volcano Tourism*) – countries (like Iceland) market their volcanoes to attract tourists – *The Land of Fire and Ice* • **Geothermal Power** – Hot rocks can be utilised for Geothermal Power (see <u>Energy 3.5</u>), a renewable energy resource • **Minerals / Precious Metals** are sometimes found in areas of volcanic activity. This is because they are transported nearer to the Earth's surface by tectonic motion, allowing them to be mined and exploited • **Sulphur Deposits** found in some volcanic craters is useful in making medicines, weapons and paper

2.1 EARTHQUAKES AND VOLCANOES

Reducing Impacts

Explain what can be done to reduce the impacts of . . . volcanoes

You can't do anything to stop a volcanic eruption from happening; but with modern technology it is becoming increasingly easier to prepare for them by **prediction**. Warning signs are visible before a major eruption, examples including:

- Steam and Gas being emitted – detected sensors placed next to dormant or active volcanoes
- Lava Flows
- Ash emitted – can be seen from ground or air
- Bulges – can identified from satellite imagery
- Minor earthquakes from magma moving into the volcano
- Changes in composition of the air surrounding the volcano – like increased amounts of sulphur or CO_2

The major problem is that people don't want to evacuate until the threat is eminent – because the prediction of volcanic eruptions isn't completely perfect. This makes them more liable to becoming injured or killed by the event.

Image 17: A Volcanologist on Etna

Volcanologists are responsible for studying volcanoes – meaning that they have to make predictions. They use modern technological developments to better predict eruptions. Better prediction constitutes to better readiness.

People can also be prepared by staying indoors with all doors and windows closed (to prevent ash entering the house / lungs) with sufficient food and water supplies during an eruption. People also need to be careful when the air is thick with volcanic ash – because it could potentially be harmful when inhaled.

THEME 2
THE NATURAL ENVIRONMENT

Case Study – Eyjafjallajökull, Iceland
A Volcano

Image 18: Fissure on Fimmvörðuháls (part of the Eyjafjallajökull system)

Eyjafjallajökull Visitor Centre, Iceland

Location

Iceland lies on the Mid-Atlantic Ridge, a constructive plate boundary between the Eurasian and North American plates. Because the 2 plates move apart, magma rises to the surface, forming active volcanoes in diagonal belt across Iceland (SW to NE).

Eyjafjallajökull is located beneath an ice cap in the south of the island, named so because of the glacier it sits beneath.

Eruption March 2010

In March 2010, magma superseded the crust beneath the Eyjafjallajökull glacier. This initiated 2 months of powerful eruptions. In March, the early eruptions were lava based – meaning they posed less of a threat to people living nearby.

On the 14th April, a more explosive series of eruptions began.

Over just a few days, violent eruptions propelled massive amounts of ash into the atmosphere.

Local Impacts and Responses

The heavy particles of ash fell close to the volcano, forcing hundreds of people from their farms and settlements (this was an immediate response). Rescuers had to wear face masks to stop them from choking.

The ash falling on the fields, devastating crops, was the primary effect – because the ash came from the eruption.

The secondary effects included flooding from some melting of the glacier which the volcano was beneath (Eyjafjallajökull). This caused

2.1 EARTHQUAKES AND VOLCANOES

a torrent of meltwater to flood down onto South Iceland's lowland plains. This would have destroyed vast areas of Rt. 1 (Iceland's main road) if the embankments weren't deliberately breached – which was done to prevent damage to expensive bridges along the road. Later, the embankments could be quickly rebuilt – meaning the highway was repaired in only a few weeks.

Local Impacts

- 800 people evacuated
- Homes, Roads, Farms (crops) and services like water + electricity were damaged or disrupted
- Local water supplies were contaminated with fluoride (found in ash)

National Impacts

- Reduced tourism numbers because people thought Iceland was dangerous. This impacted Iceland's economy and reduced tourist interest – putting people's jobs at risk
- Transportation was disrupted due to some sections of road in the south being washed away by the flash flooding
- It was expensive to rebuild roads, houses and farms along with restoring services
- Agriculture was affected because the crops were covered in a thick layer of ash by the eruption

International Impacts and Responses

The eruption became an international event when the cloud of finer ash (unlike the dense ash that settled in Iceland, light fine ash can be carried long distances by wind) spread SE towards Europe. A large portion of European airspace was grounded because people were concerned about how volcanic ash would affect plane engines. The grounded planes and cancelled flights stranded people and prevented freight (like food) travelling around Europe. This meant that food perished (like beans in Kenya) as it couldn't be transported into the European markets to be sold. So, because foreign farm workers were affected and planes were grounded, the eruption had international impacts.

- 100,000 flights cancelled over 8 days
- 10 million people's flights affected
- Economic losses were expected to amount to about £80 million
- Food and goods couldn't be transported (because flights were cancelled). This meant that factories producing products had to halt manufacturing because the raw materials couldn't reach the factories

THEME 2
THE NATURAL ENVIRONMENT

Earthquakes

Earthquakes occur below the Earth's surface when rocks make large movements to release intense stress and pressure created by tectonic activity.

Features of Earthquakes

Features of earthquakes (including epicentre, focus, magnitude)

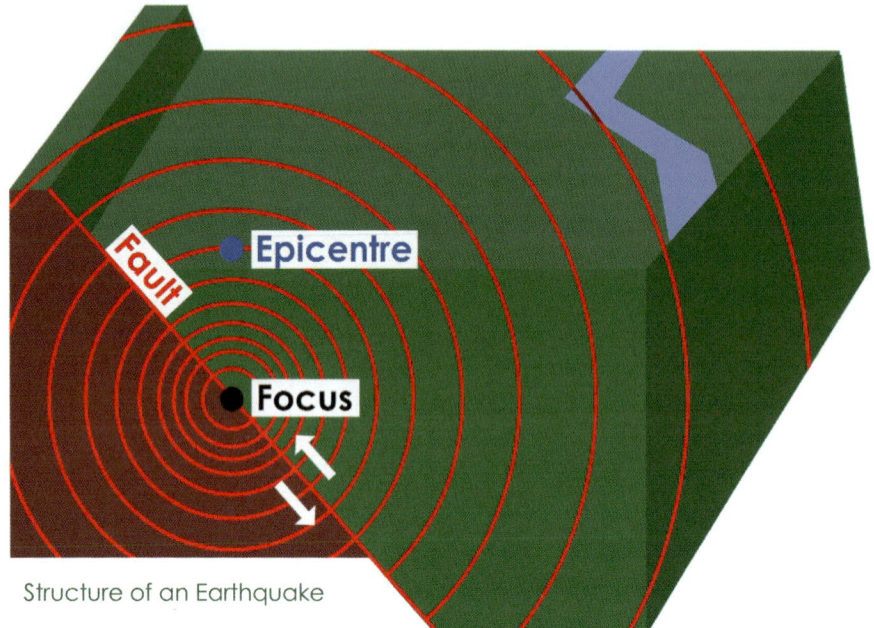

Structure of an Earthquake

The point where an earthquake originates from is called the **focus**.

The point on the surface directly above the focus is called the **epicentre**.

The **magnitude** of an earthquake is how strong it is or the scale of its effect on the environment and people.

The **Mercalli Intensity Scale** is a method of measuring earthquake intensity based on its effects and its damage.

The **Richter Scale** is a method of measuring earthquake intensity by the magnitude of an earthquake (the total amount of energy released).

Causes of Earthquakes

Describe the causes of earthquakes . . .

Earthquakes are caused by tectonic plate movements. Pressure builds up when plates get stuck and the plates continue to try to move. The pressure is created from friction between the 2 plates.

2.1 EARTHQUAKES AND VOLCANOES

Pressure is released when the plates break free along a fault line. The shockwaves from this release travel as seismic waves through the ground.

1. 2 plates try to move past each other
2. Friction between them creates pressure
3. When this pressure is released, huge pulses of energy are sent out
4. These shock waves go in all directions

NB The shallower the focus, the greater the impact of the earthquake above the ground.

Hazards of Earthquakes
Describe . . . their effects on people and the environment

	People	Environment
Positive	• Underground minerals may be brought closer to the surface by earthquakes. People can mine these for money.	No real benefits to the environment
Negative	• Death from falling plants and buildings • Destroy buildings • Destroy infrastructure • Tsunamis caused may kill people	• Tsunamis may destroy habitats • Earthquake movement may also destroy habitats

Buildings being destroyed is a primary impact. The large amount of money needed to rebuild them is a secondary impact.

The impacts of earthquakes are in some ways greater in MEDCs because buildings and infrastructure are built using more expensive materials – meaning it is costlier to repair them than in LEDCs. Buildings are built with cheaper materials in LEDCs, meaning that they are cheaper to repair. But, in LEDCs, buildings are less well constructed – meaning that they are more likely to collapse and create a negative effect on the population.

Factors that affect the damage an earthquake causes include:

- The amount of energy released (more energy means more damage)
- The depth of the focus (a deeper focus means there is less damage)
- The density of population near epicentre
- Whether the buildings are earthquake-proof

THEME 2
THE NATURAL ENVIRONMENT

- How solid the ground / bedrock is (if it is more solid, less energy will be absorbed, and the damage will be greater)

Reducing Impacts
Explain what can be done to reduce the impacts of earthquakes . . .

Ways to reduce the impact during an earthquake:

- Keeping Calm
- Staying indoors
- Taking cover in a doorway or under a table
- Keeping away from windows and heavy furniture

Seismic Design reduces the impact of earthquakes, as it reduces the likelihood of buildings collapsing – meaning that less buildings need to be rebuilt. It does this because it allows buildings to absorb the earthquake's energy or makes them strong enough to remain standing during an earthquake.

Some examples of seismic design:

- Motion Dampeners
- The ability for the building to sway (stabilised by a central pendulum), or by using bendy materials to construct the building
- Shock absorbers absorb the earthquake's energy
- Shear walls (braced panels to help to absorb sideways movement)
- Shorter buildings are less likely to collapse from strong **lateral motion**
- Concrete columns provide increased strength
- Light walls are less likely to fall off a building
- Strong building frames and foundations give a building more strength

Also, people (in earthquake-prone areas) should have an earthquake box in their house containing a radio, torch, food, clothing and water. This would keep to keep people alive and able to call for help if they were trapped in their house after an earthquake.

Case Study – Kobe Earthquake, Japan
An Earthquake

Fact file

5:46 am 17th January 1995   20 seconds long

7.2 magnitude   18cm horizontally, 15 cm vertically

2.1 EARTHQUAKES AND VOLCANOES

Causes

Japan is located on a minor fault called the **Nojima Fault**. This is where the Eurasian, Philippine and Pacific plates meet. The Philippine plate (oceanic) is forced downwards (**subducted**) beneath the Eurasian plate (continental) at a destructive plate boundary. This formed the Japanese island arc and leaves Japan prone to experiencing severe earthquakes.

Primary Effects

- Nearly 200,000 buildings collapsed
- 1km stretch of Hanshin Expressway collapsed
- Numerous bridges collapsed along a bullet train route – several trains were derailed
- 120 of the 150 quays in Kobe Port were destroyed

Secondary Effects

- Electricity, gas and water supplies were disrupted
- Fires caused by gas leaks mixing with exposed and live power lines destroyed 7500 houses (which were wooden)
- 230,000 were made homeless and had to live in temporary shelters with cold temperatures, food shortages, water shortages and few blankets
- People were afraid to return home due to over 74 strong aftershocks
- Industries (e.g. Mitsubishi and Panasonic) were forced to temporarily close

5500 deaths

40,000 injuries

180 000 destroyed houses

However, the new airport (Kansai International) and the Akashi bridge were undamaged because they were built using seismic design principles.

Image 19: Kobe Port Memorial Park

The Months After

- By July, infrastructure was fixed
- Rubble cleared
- Rail services ran in August
- The port was fixed a year later, but not the motorway
- New earthquake building standards were introduced
- More seismic equipment became available

THEME 2
THE NATURAL ENVIRONMENT

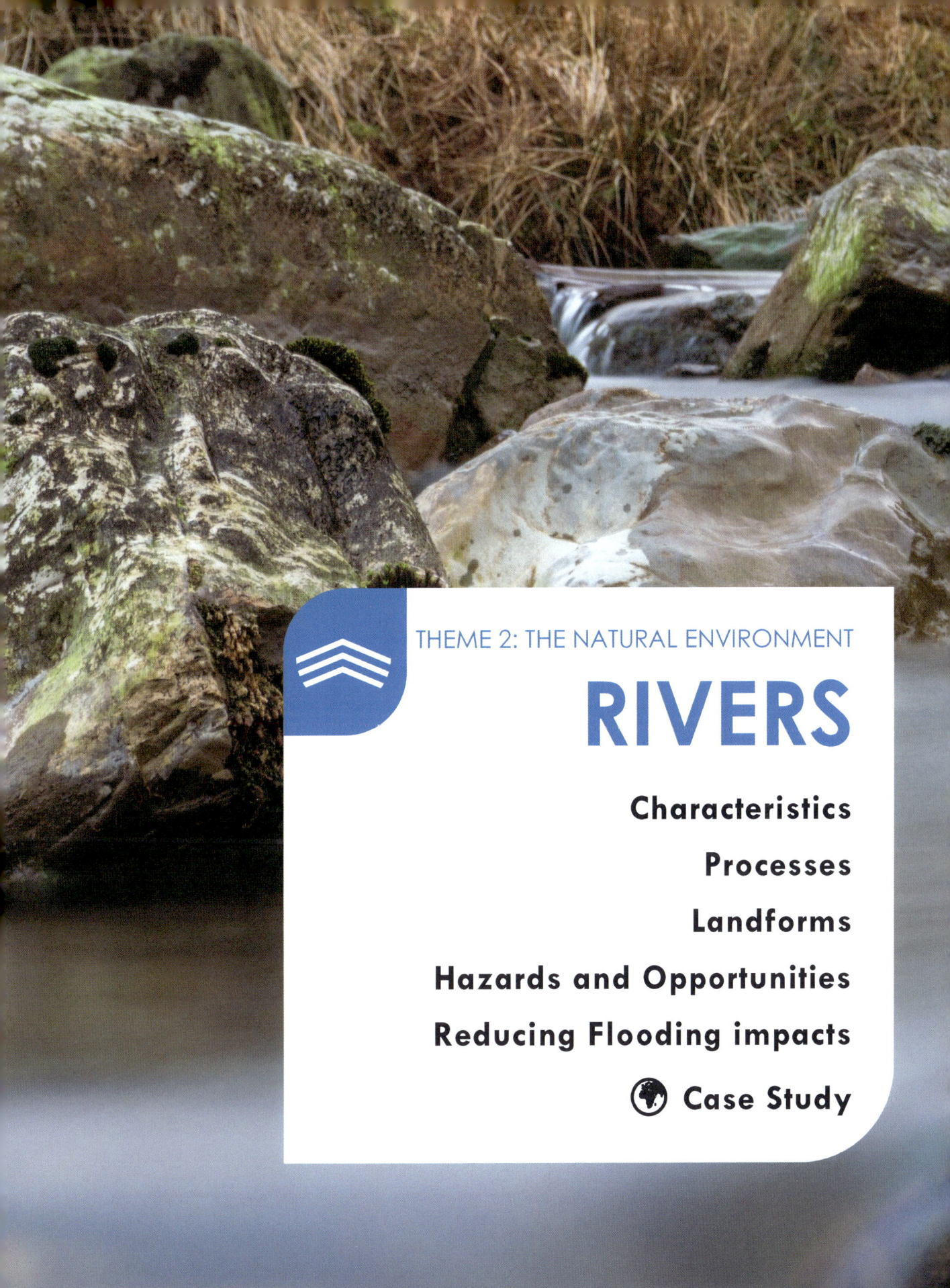

THEME 2: THE NATURAL ENVIRONMENT

RIVERS

Characteristics

Processes

Landforms

Hazards and Opportunities

Reducing Flooding impacts

🌍 Case Study

THEME 2
THE NATURAL ENVIRONMENT

Explain the main hydrological characteristics and processes which operate in rivers and drainage basins

Demonstrate an understanding of the work of a river in eroding, transporting and depositing

Describe and explain the formation of the landforms associated with these processes

Demonstrate an understanding that rivers present hazards and offer opportunities for people

Explain what can be done to manage the impacts of river flooding

Drainage Basins

Explain the main hydrological characteristics and processes which operate in . . . drainage basins

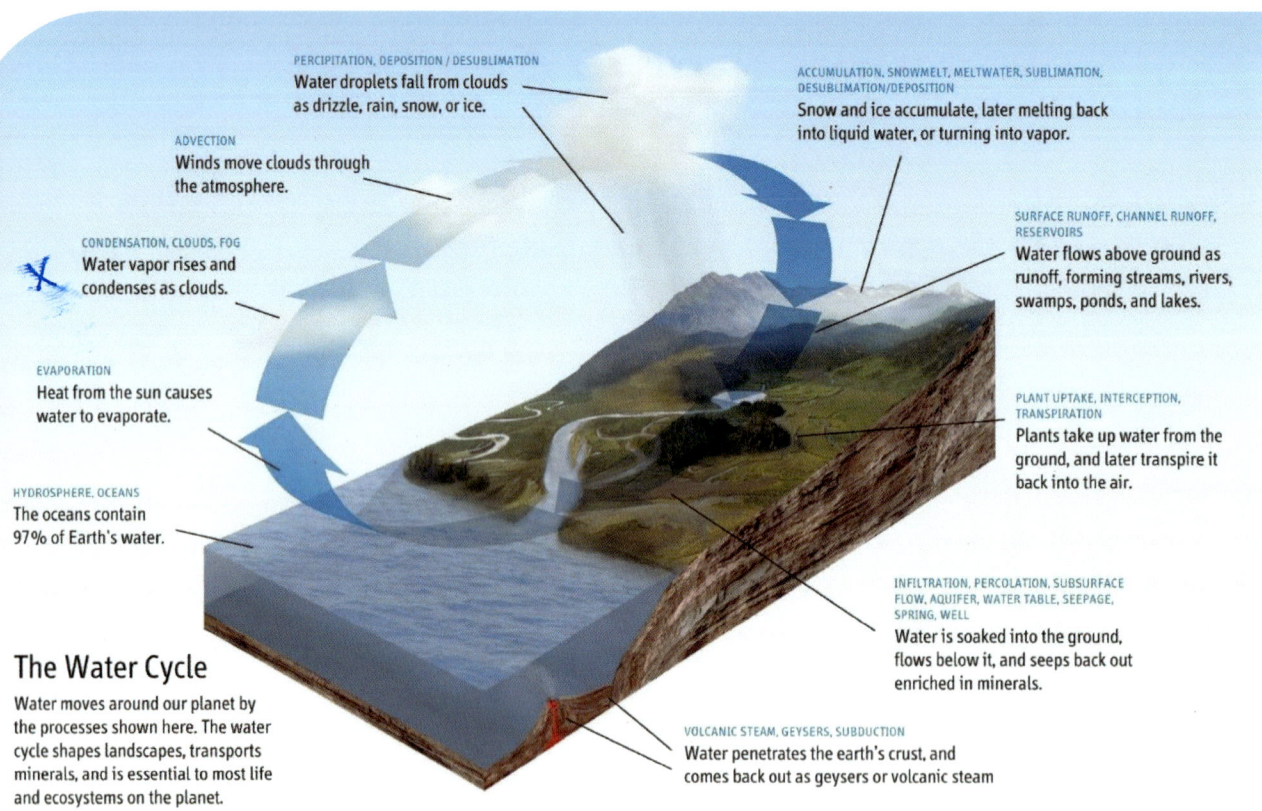

Image 20: The Hydrological Cycle

The **hydrological cycle** (also known as the water cycle) is the movement of water through the land, **hydrosphere** and **atmosphere**. It is a **closed system** as water isn't added or lost from the system.

2.2 RIVERS

Characteristics

Characteristics of . . . drainage basins (including watershed, tributary, confluence)

A **drainage basin** is the area of land from which the water content is drained into to a river or its tributaries. A river collects the runoff from this land, draining it. It's basically the river's catchment area. The slice of land shown in the previous diagram is a drainage basin for the main river in the centre. This is an open system because water can be added or removed.

The **Watershed** is the edge of the drainage basin. This is usually the ridge of hills surrounding a drainage basin. It is also the boundary between 2 different drainage basins.

A **Tributary** is a smaller river or stream that flows into a larger river or lake.

The **Confluence** is where 2 rivers join or meet.

Processes

Processes which operate in a drainage basin (including interception, infiltration, throughflow, groundwater flow, evaporation, overland flow)

Inputs – water added	Transfers – water moving	Stores – water not moving	Outputs – water leaving
• **Precipitation** is water falling in any form from the sky	• **Overland Flow** is when water travels across the surface of the ground • **Channel Flow** is when water travels in channels (e.g. rivers) • **Infiltration** is when water travels from the surface (surface storage) into the ground • **Throughflow** is the movement of water through *unsaturated ground*	• **Interception** is when an object (like a tree) stops the precipitation from reaching the ground • **Surface Storage** is water held on the surface of the ground – like in a lake • **Soil Moisture Storage** is water stored in pores in *unsaturated ground*	• **Evaporation** is when liquid water in surface stores becomes a gas • **Transpiration** Is when liquid water evaporates from vegetation (like trees) • **River Discharge** is when the water carried in a river is released into the sea

77

THEME 2
THE NATURAL ENVIRONMENT

Inputs – water added	Transfers – water moving	Stores – water not moving	Outputs – water leaving
	• **Groundwater Flow** is the movement of water through *saturated ground*	• **Groundwater Storage** is water stored in *saturated ground*. It makes the ground saturated, forming the water table	

Saturated ground is rock or soil where all the pores are filled with water because they are **beneath** the **groundwater table**. Water flowing through this ground is called **Groundwater Flow**.

Unsaturated ground is the rock or soil **above** the **groundwater table**. The soil and rock may contain air in their pores. Water flowing through this ground is called **Throughflow**.

Gulfoss Waterfall, Iceland

River Characteristics

Explain the main hydrological characteristics and processes which operate in rivers . . .

Characteristics of rivers (including width, depth, speed of flow, discharge)

- The **Width** of a river is how wide it is (from bank to bank)
- The **Depth** of a river is how deep it is (from surface to riverbed)
- The **Speed of Flow** (or **Velocity**) of a river is how fast the river water is moving. It increases as the cross-sectional area increases (because less water has contact with the banks and bed, meaning there is reduced friction) and as the load decreases (because there is less resistance and friction from big stones)

78

- The **Discharge** of a river is the amount of water flowing in a river / channel (Cross Sectional Area * velocity)
- The **Cross-Sectional area** is the area of a slice of flowing water (the area of a cross-section of the river)
- The **Long Profile** of a river is how a river's gradient changes over the course of a river
- The **Bradshaw Model** is a model used to show how characteristics of a river change from the source to the mouth

River Processes

Demonstrate an understanding of the work of a river in eroding, transporting and depositing

Erosion

Transported material (the load) is used by a river for erosion. As the velocity of a river increases, so does the amount of load it can carry, and the amount of erosion it can carry out. Rivers erode in 4 main ways:

- **Hydraulic Action** – is erosion caused by the force and impact of flowing water. This force and impact removes material from the bed and banks of a river
- **Corrasion (Abrasion)** – is the erosion caused by the load rubbing against the bed and banks of a river, scraping off material. This is possibly the biggest cause of erosion in a river
- **Attrition** – the process of the river's load being eroded as it wears away the bed and banks. This makes the bedload rounder over the course of the river
- **Corrosion (Solution)** – the dissolving of rocks. Limestone dissolves in water, and the river carries this material away – inducing erosion

Transportation

Eroded materials transported by the river are called the **load**. There are 4 main types of load (in order of energy requirement)

- **Solution** – minerals are dissolved in the river's water (LEAST ENERGY)
- **Suspension** – light materials are carried near the surface of the river, tinting the river. When soil is carried, a murky brown colour is created
- **Saltation** – small pebbles and stones transported along the riverbed – perhaps bouncing
- **Traction** – large, heavy boulders and rocks are rolled along the riverbed (MOST ENERGY)

79

THEME 2
THE NATURAL ENVIRONMENT

Deposition

Material needs energy to be transported. So, when a river has less energy (like in the lower course), the river deposits material. A river usually has less energy (and velocity) when the gradient is shallow, and has more when the gradient is steep. Heavy materials are deposited first, and fine materials last.

Fine materials sometimes aren't deposited, so when a river floods they are left behind when the water is reabsorbed back into the river. This material is fertile and can build up over multiple flood events to create a fertile flood plain.

Eroding River Valleys

- **Vertical Erosion** – Downwards erosion
  - As a stream or river flows downhill, gravity encourages the water to reach sea level. This means the river bed is eroded, and the river channel gets deeper. This smooths out the river's long profile by evening out uneven gradients.
- **Lateral Erosion** – Sideways erosion
  - The river erodes at its banks, widening the river channel. Weathering can also cause lateral erosion due to particles being removed by exposure to strong winds and precipitation.

River Landforms

Describe and explain the formation of the landforms associated with these processes

Forms of river valleys – long profile and shape in cross section, waterfalls, potholes, meanders, oxbow lakes, deltas, levées and flood plains

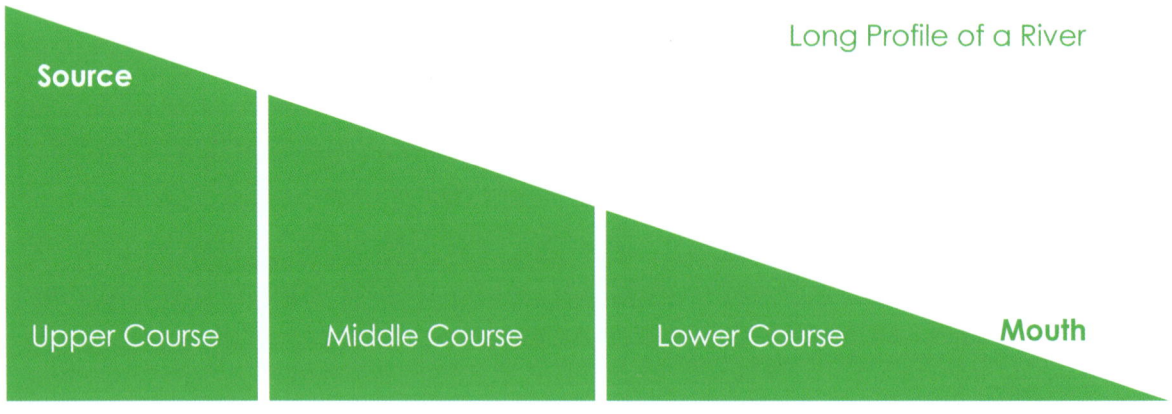

Long Profile of a River

2.2 RIVERS

Upper Course

Near the source, the river has a steep-sided V-shaped valley, steep gradient and a river channel occupying most of the valley floor.

The river also carries a large load because the transported material hasn't had enough time to be eroded down by **attrition**. The load found on the riverbed is called **bedload**. There is lots of friction between the water and the large bedload, meaning that the river has a low velocity.

The large amount of vertical erosion (because of uneven gradients) in this section of the river create steep-sided V-shaped valleys, rapids, waterfalls and gorges.

High vertical and low lateral erosion creates the distinctive V-shaped valley.

Potholes

If the river bed is uneven, pebbles bouncing along in **saltation** can become stuck in groves. Swirling currents make the pebbles rotate and erode circular holes or grooves in the riverbed.

Rapids

Alternating bands of hard and soft rock across the river's flow form an uneven bed (because the soft rock is more easily eroded), which leads to areas of more turbulent water called **rapids**.

Waterfalls

- Form where a horizontal layer of hard rock lies above a layer of soft rock, like in river valleys
- The soft rock underneath is eroded faster than the hard rock on top
- Over time, a plunge pool develops
- The splashing water, combined with eddy currents in the plunge pool undercuts the hard rock shelf
- This produces an unsupported overhang that then collapses
- Repetition of this undercutting and collapse means that the waterfall will retreat upstream, creating a deep, steep-sided valley called a **gorge**

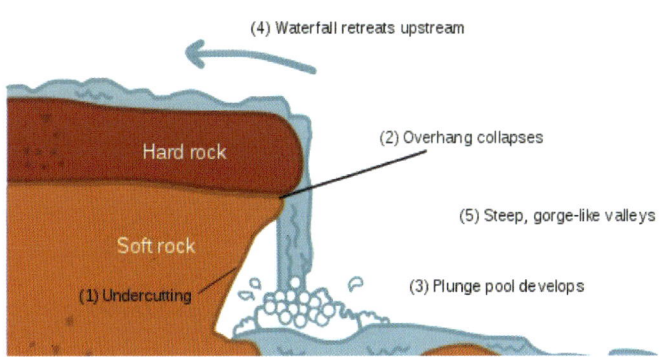

Figure 9: Waterfall Formation

THEME 2
THE NATURAL ENVIRONMENT

Middle Course

In the middle course of a river, the river channel is wide and deep with gentle valley sides and a wide floodplain. The river carries more water because tributaries have joined it at points of confluence.

Lateral erosion becomes more prominent than vertical erosion (as there is an increased velocity with less abrupt gradient changes) – meaning the river starts to bend or meander across the flood plain. Erosion in the upper course has decreased the bedload size, and there is increased deposition because the river has less energy.

Meanders ✓

The velocity of a river is different across its river channel, so erosion is different. There is more erosion on the side with a greater velocity, meaning that the river cuts into the riverbank, making the river wider.

There is more deposition on the side of the river with a lower velocity. This is because a lower velocity means the river has less energy, increasing deposition. This deposition forms a 'slip-off slop', narrowing the river. A **meander bend** is formed. The meanders keep their shape as they erode sideways across the flood plain, due to these 2 processes.

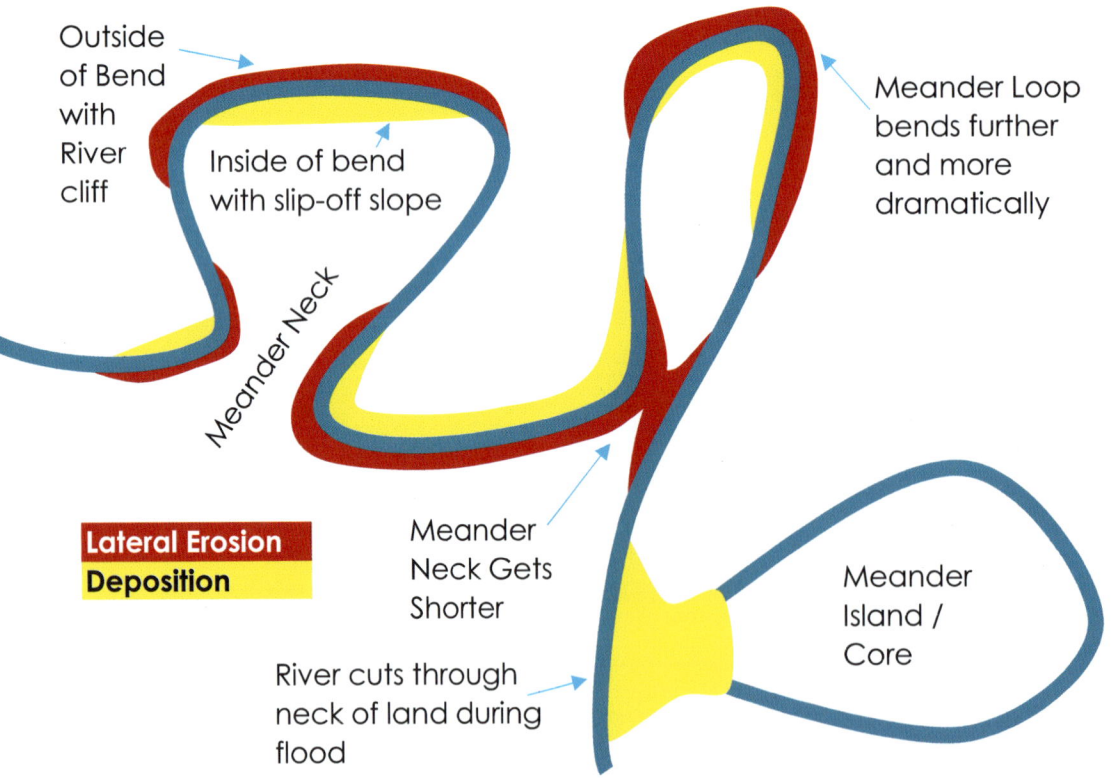

Structure of Meander Bends, and the Formation of Oxbow Lakes

2.2 RIVERS

Oxbow Lakes

As the loop of a meander gets bigger, the neck of the bend gets narrower, becoming a 'swan neck' meander.

In times of flood, the river might erode a fresh channel to straighten its course (note the new channel in the previous diagram). This completely cuts off the meander neck.

The old meander loop becomes an oxbow lake when its ends are sealed by deposition (the yellow). Over time, the **oxbow lake** may dry up and become filled with vegetation.

Image 21: - Mississippi River South of Muscatine, Iowa, USA

Lower Course

In the lower course of a river (near the mouth), the river is broad and deep. The bedload is transported in suspension or solution and there is increased deposition (because the river has less energy, due to a shallower gradient). The river has a high velocity because there is little friction with the bed, banks and load with less big stones.

Flood Plains and Levées

A river meanders across its flood plain.

The **flood plain** is the area either side of the channel which floods when the river bursts its banks. Also, the river channel's edges have natural ridges called **levées**.

When the river floods, the largest particles are deposited first (building up the levées with each flood). Finer material (alluvium) is spread out in layers over the rest of the flatter flood plain. This material is lighter, so it travels further.

Deltas

The form of the river channel at the river's mouth depends on the rock type, strength of waves, tides or load. The river may end in a wide, deep estuary or a narrow mouth. In most rivers, a delta formation will mark the end of a river's long profile.

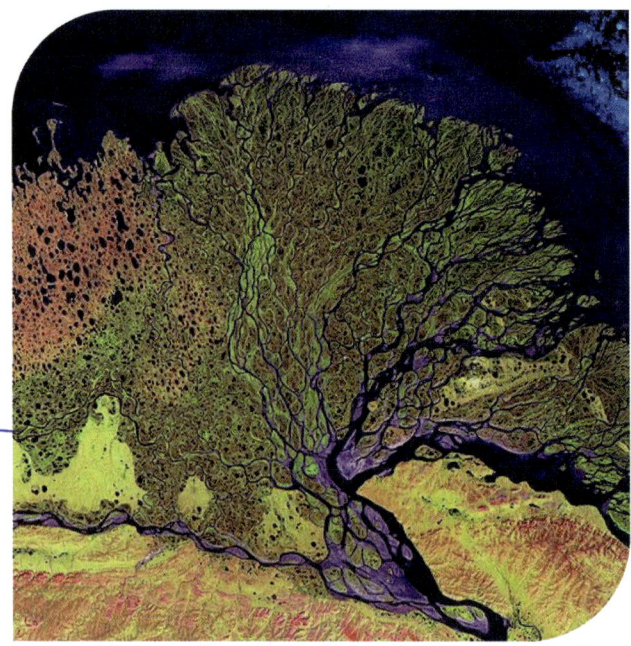

Image 22: Lena River Delta False Colour - (Landsat)

Deltas form when rivers carry lots of sediment:

1. As the river reaches the sea, the velocity of the river dramatically reduces – creating lots of deposition (as the river has less energy)
2. As sediment is deposited, it blocks the water flow
3. The river then divides up into separate channels (called **distributaries**)
4. Deposition continues, forming a delta that builds outwards over subsequent flow blockages

2.2 RIVERS

Cross Sections + Summary

- **Upper Course** – Steep Sides, V-Shape Valley
  - Small cross-sectional area, but large bank and bed to water ratio – creating high friction and a low velocity
  - Lots of bedload – leads to a low velocity
  - Uneven gradients produce high vertical erosion
  - Steep gradient overall means the river has **high energy**
- **Middle Course** – Gentle Slopes, Flood Plain
  - Larger cross-sectional area and smaller bank and bed to water ratio – meaning there is less friction with a greater velocity
  - Smaller, smoother bedload – producing a higher velocity (as there is less friction with the bedload)
  - Smooth gradients mean less vertical and more lateral erosion
- **Lower Course** – Very flat flood plain, Levées, Delta
  - Larger cross-sectional area and a very small bank and bed to water ratio, meaning there is less friction and a greater velocity
  - Very shallow gradient means the river has **low energy** with more deposition

Hazards and Opportunities

Demonstrate an understanding that rivers present hazards and offer opportunities for people

Causes of hazards (including flooding and river erosion)

Opportunities of living on a flood plain or a delta or near a river

Hazards	Opportunities

- **Flooding**
  - Loss of Houses and Buildings
  - Stress, Injuries and Deaths
  - Higher House Insurance
  - Loss of Transport Routes
  - Loss of Services and Infrastructure
- **Erosion**
  - Collapse of river banks (causing destruction of infrastructure and buildings)
  - Greater risk of flooding
	- Water Availability
	  - Domestic
	  - Industry
	  - Agriculture
	- Transport (boats)
	- Hydroelectric Power
	- Leisure and Tourism potential
	- Flat flood plains
	  - Good for building
	  - Fertile soils
	  - Easy to build infrastructure
	- Deltas have fertile soils for agriculture

THEME 2
THE NATURAL ENVIRONMENT

So, what's the hazard?

Floods happen when rivers don't have the capacity to carry extra discharge (this discharge is created by events):

- Storm Surges
- Heavy Rainfall
- Dam Failure

In times of rapid flow, like after heavy rain, rivers erode banks. Over time, a river builds up its banks and bed by deposition – raising the water level of the river above the ground level of the surrounding flood plain, posing a flood risk.

If the river discharge increases (like after heavy rain), the river can breach the levées it created and flood a larger area than if no deposition occurred (the water level is already above ground level as a result of deposition raising the river's water level).

Rivers may undercut banks, causing a collapse. This could destroy infrastructure and buildings.

Reducing the Impacts
Explain what can be done to manage the impacts of river flooding

Flood prevention helps to stop floods from occurring. Hard and Soft engineering techniques could be used:

- **Hard Engineering Flood Prevention** – expensive, permanent structures that affect the flow of rivers, preventing flooding
    - **Dams** – can hold back water in reservoirs, meaning the water can be released slowly over time (however, flooding land to form a reservoir destroys habitats)
    - **Flood Walls** – can hold back water in flood events (however, they look ugly)
    - **Embankments** (levées) – increase the capacity of the channel and reduce the chance of overflow. Often, they are made from concrete or stone to reduce their erosion. (But they are ugly, and could cause more severe flood events – as the river's capacity is increased, more water can flood an area than before)
    - **Dredging** – making a channel deeper increases its capacity and reduces the chance of overflow (however, dredging can cause flooding downstream)

2.2 RIVERS

- o **Bridge Design** – Bridges with wide pillars and walls act like dams during floods, meaning water is released more slowly so it doesn't overwhelm the river's capacity. However, most modern bridges aren't like this
- **Soft Engineering Flood Prevention**
  - o **Afforestation** – Planting trees in a drainage basin means that vegetation intercepts rainwater, releasing it by transpiration into the air. This means that less water will reach the river at once, reducing the chance of it flooding
  - o **Wash lands** – water is allowed to flood into unimportant land (like unused fields) so it doesn't flood important land (like settlements)
  - o **Land-use restrictions** – only certain buildings are allowed to be built next to rivers. For instance, recreational areas could be permitted near rivers with housing and important services further, and higher, than the river

Image 23: Hoover Dam, Arizona, USA

THEME 2
THE NATURAL ENVIRONMENT

Case Study – Mekong River, SE Asia
The opportunities presented by a river or rivers, the associated hazards and their management

The Mekong river is one of the world's biggest river systems. It flows for nearly 5000km through 6 countries: China, Myanmar, Thailand, Laos, Cambodia and Vietnam.

The reason why the river is so important is because of its seasonal variations which lead to natural, seasonal floods. These help to fertilise crops and provide habitats for many organisms.

Opportunities

- **Biodiversity**
    - The Mangroves in the Mekong delta provide habitats for animals
    - Mekong fish species depend on the annual flood. They migrate to deep pools in dry seasons and to the nutrient-rich floodplains for feeding and spawning
- Agriculture
    - Mangroves in the delta prevent erosion and trap nutrients, which is useful for farming crops
    - Fertile land means that crops have higher yields
    - Sediment in the delta is very good for farming and, farms here produce around 90% of Vietnam's rice
    - Sugar Cane grows in the delta
- Fishing – The Mekong fish thrive due to the annual floods
- Natural Resources – The flood plains are mineral rich
- Transport – The river makes it easy to transport and sell goods
- Tourism in the Mekong Delta
    - People go to the Mekong Delta to see the stunning mangrove swamps, markets and delta river towns
    - Buddhist temples
    - Birdlife and wildlife
- Housing – Many people choose to live on houseboats on the water – this means they don't flood, and that they can be moved when they change farming pastures
- Water – many people along the Mekong rely on the river for drinking, washing, cleaning and transport. It is an essential method of transportation and as a mechanism for trade in the region. 73% of cargo in Vietnam is transported on the Mekong.

2.2 RIVERS

Flooding

There is an annual flooding which can cause damage if people aren't prepared for it – for instance, it could destroy crops.

Severe flooding can result in loss of life, damage to buildings and infrastructure, damage to agriculture and disruption of human activities. **Cambodia and Vietnam** make up **two thirds** of the Mekong Basin's flood damage each year.

Flash floods are the largest risk though because people often don't have access to communication mechanisms to allow them to prepare. This lack of preparation destroys houses and infrastructure.

Also, as **people rely on the river** so much, **drought** is a big problem. It can result in food and water shortages, loss of money and damage to agriculture.

Flood Management

- Forecasting – During the flood season, daily flooding forecasts are issued. These give communities in Laos and Cambodia a chance to prepare for floods
- Flood protection **dikes** and control gates to control water in flash floods. These help to reduce the impact of unexpected flash floods
- Dikes to prevent the flooding of **Vientiane**, the capital of Laos
- Protecting the river banks by placing rocks on the sides of the rivers to prevent erosion of river banks

Image 24: Mekong River at Luang Prabang, Laos

THEME 2
THE NATURAL ENVIRONMENT

THEME 2: THE NATURAL ENVIRONMENT

COASTS

Processes
Landforms
Coral Reefs
Mangrove Swamps
Hazards and Opportunities
Reducing Impacts

🌍 Case Study

THEME 2
THE NATURAL ENVIRONMENT

Demonstrate an understanding of the work of the sea and wind in eroding, transporting and depositing

Describe and explain the formation of the landforms associated with these processes

Cliffs, wave-cut platforms, caves, arches, stacks, bay and headland coastlines, beaches, spits, and coastal sand dunes

Describe coral reefs and mangrove swamps and the conditions required for their development

Demonstrate an understanding that coasts present hazards and offer opportunities for people

Hazards (including coastal erosion and tropical storms)

Explain what can be done to manage the impacts of coastal erosion

The **coast** is the place where the land meets the sea.

The **coastline** is the outline of the edge of the land on a map.

Waves result from friction between the surface of the sea and the wind. This makes part of the sea rise at right angles to the wind.

The size of a wave depends on the:

- Wind speed
- Time which the wind blows in the same direction
- Fetch of sea (length of sea which the wind blows over)

There are 2 types of waves:

Constructive Waves carry out deposition – creating a wider, more gently sloping beach. Longshore drift can create sand dunes on these beaches. They have a long wavelength and are low, with a weak backwash and strong swash. Deposition occurs because material is brought up onto the shore by the strong swash and isn't removed again by the weak backwash.

Destructive Waves carry out erosion and take material off beaches – creating a steep, narrow beach. They have a short wavelength and are tall, with a strong backwash and weak swash. The beach is eroded because material is removed from it by the very strong backwash.

Waves at Reynisfjara, Iceland

2.3 COASTS

Constructive Waves	Destructive Waves
Deposition	Erosion
Long Wavelength	Short Wavelength
Low, shallow waves	Tall, steep waves
Strong Swash	Weak Swash
Weak Backwash	Strong Backwash
Wide beach with more material	Narrow Beach with little material

Coastal Processes

Demonstrate an understanding of the work of the sea and wind in eroding, transporting and depositing

Marine Erosion

1. **Hydraulic Action** – water is forced into cracks in the rocks, compressing the air inside. When the wave retreats from the rock, this compressed air is released. It can force the rock apart if the pressure is high or the rock soft or fragile
2. **Corrasion (Abrasion)** – sediment is thrown against the cliff by waves. It wears the cliff-face away and chips off bits of rock
3. **Attrition** – loose sediment that is knocked off the cliff by hydraulic action or abrasion is moved and swirled around by the waves. It collides with other sediment, wearing the sediment down into smaller and rounder particles
4. **Corrosion (Solution)** – when seawater dissolves material from the rock. This is common on limestone and chalk coasts where the calcium carbonate can be dissolved in the sea

Marine Erosion

The processes of marine erosion are like those of river erosion.

The effects of attrition increase the further the distance you are from the cliff and the longer the period they are moved by the waves for. A boulder, over time, will break down into grains of sand which can't be broken down any further.

Rounded beach material, between the size of boulders and sand, is called **shingle**.

THEME 2
THE NATURAL ENVIRONMENT

Transportation

Marine Transportation

1. **Suspension** – Fine sediment is carried as a suspension in the water, creating a murky appearance
2. **Traction** – Larger Pebbles are rolled along the seabed
3. **Solution** – Dissolved material is carried invisibly in solution in the water
4. **Saltation** – Small pebbles are moved by collisions. One pebble hits another, making it bounce. This can set up a chain reaction of colliding and moving pebbles

The sea transports the load and sediment from erosion in the same way rivers do.

Sediment can be moved up and down a beach by waves, but also along it by **longshore drift**.

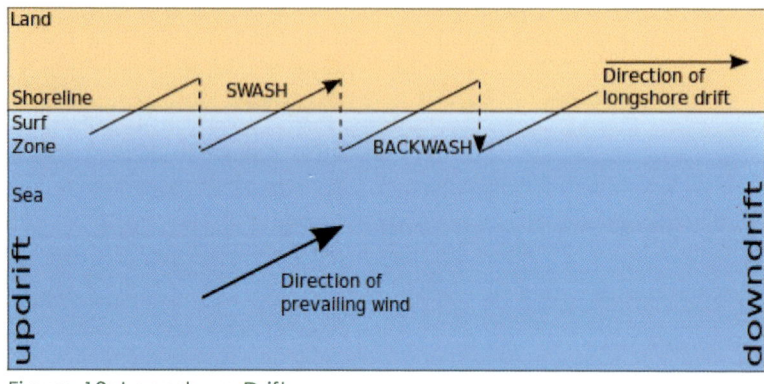

Figure 10: Longshore Drift

Longshore Drift

1. Prevailing winds mean that waves approach the shore at an oblique angle
2. The swash of a wave carries the pebble at an oblique angle up the beach
3. The pebble is pulled back down the beach at a 90-degree angle by the backwash

Repetition of this process over time moves the pebble along the beach, in a **zig-zag motion**, in the lateral (sideways) direction of the prevailing winds.

2.3 COASTS

Deposition

Material is taken off a beach by **destructive waves** and is deposited by **constructive waves**. Sand and pebbles build up to form more land at a coast.

Coastal Landforms

Describe and explain the formation of the landforms associated with these processes

Cliffs, wave-cut platforms, caves, arches, stacks, bay and headland coastlines, beaches, spits, and coastal sand dunes

Coastal Erosion
Why is coastal erosion more rapid on some coasts than others?

- The main rock type – whether the rock is affected by corrosion or not
- Presence of coastal defences – these slow down erosional processes
- Whether the rock type is soft or hard (soft rock erodes faster)
- The strength of the coast's waves and prevailing winds (if stronger, erosion is more rapid)
- If the coast is sheltered – reduces the wind's weathering and erosion

Cliffs and Wave-Cut Platforms

1. Weathering weakens the rock at the top of a cliff
2. The sea attacks the base of the cliff (hydraulic action), creating a wave-cut notch
3. This notch increases in size, meaning the cliff collapses
4. Backwash carries the collapsed material towards the sea, forming a wave-cut platform. This platform is at the level of the old wave-cut notch
5. Over time, the cliff retreats

Cliffs are vertical slabs of rocks formed by weathering and marine erosion.

Wave-Cut Platforms are narrow, flat areas found at the bottom of a cliff. Created by the wave's erosion and the collapse of the ceiling of a wave-cut notch.

Image 26: Wavecut Platform

Image 25: Cliffs of Bonifacio

THEME 2
THE NATURAL ENVIRONMENT

Caves, Arches, Stacks

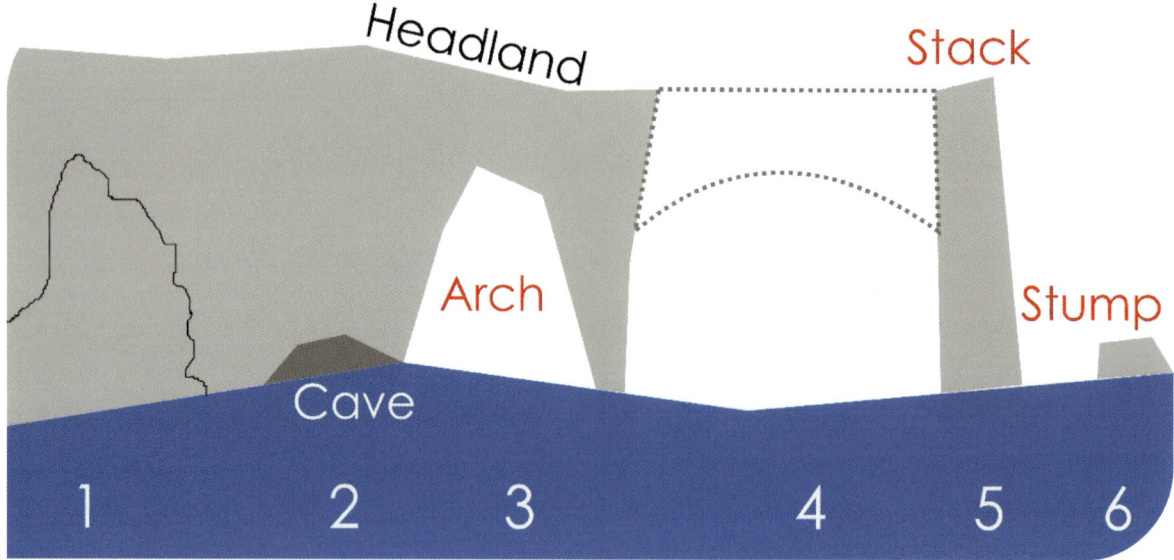

The Formation of Caves, Arches, Stacks and Stumps

1. Fault in Resistant Rock
2. Corrasion and hydraulic action widen the fault to form caves on each side of the headland
3. These caves are eroded further until they cut completely through the headland and meet to form an arch.
4. The arch is eroded and the roof becomes too heavy for the weak arch to hold, so it collapses
5. This leaves a tall stack
6. The stack is weathered and eventually collapses, leaving a stump

Reynisfjara, Iceland

2.3 COASTS

Bays and Headlands

These are formed on coasts with alternating bands of hard and soft rock.

The bands of soft rock (like clay or sandstone) erode faster than the bands of hard rock (like chalk). These means that over time a section of land is left jutting out into the sea. This is called a **headland**. The areas between headlands, where the soft rock has been eroded away, are called **bays**.

Discordant Coastlines have more bays and headlands, as there are alternating rock strata of hard and soft rock. **Concordant Coastlines** have less bays and headlands as they are made up of the same type of rock.

Coastal Deposition

Coastal Deposition is the process involving the sea's load being dropped or left on a beach by the waves.

Features created by coastal deposition are often a result of longshore drift. Most beach material is transported along coasts by this process.

Beaches

Beaches are the result of deposition by constructive waves.

They are made up of material that has been eroded, transported and deposited by the sea.

The smallest material on a beach is next to the water (as waves break more frequently here, causing a higher amount of attrition).

The largest material on a breach is at the back, near the cliffs. This material is deposited when the sea has a lot of energy.

Sandy beaches tend to have gentler profiles, and shingle beaches steeper profiles. The material found on a beach depends on the wave energy of the coast and the geology of the area.

THEME 2
THE NATURAL ENVIRONMENT

Image 27: Bruny Island Spit

Spits

A **spit** is a long, low and narrow ridge of sand or shingle that has one end attached to the coast, with the other end in the sea. They are formed by coastal deposition.

1. Sediment is carried along the coast by Longshore Drift, following the direction of sea currents and the prevailing wind
2. The material moved in this process is deposited when the coast suddenly changes direction
3. As the deposition continues, material builds up on the sea bed, forming a spit over time

NB There may be a hook shape at the end of a spit, formed by secondary prevailing winds. Also, dunes can also form on the spit. Often, a salt marsh forms behind the spit.

Sand Dunes

1. Sometimes found at the top of beaches
2. At low tide, deposits of sand on the beach will begin to dry out
3. When the wind blows towards the land, the sand will be transported up the beach
4. At the top of the beach, this sand can become lodged by wood
5. Over time, the lodged sand builds up, forming a sand dune
6. **Marram grass** often grows on newly-formed dunes, its long roots binding the sand together

98

2.3 COASTS

Coral Reefs
Describe coral reefs … and the conditions required for their development

Coral Reefs are made up tiny marine animals called **Coral Polyps** that live in colonies. Their skeletons are calcareous and join up with others to form a hard mass.

As 1 generation of polyps dies, the next generation grows on top, growing the reef.

A reef needs a solid surface to grow on. This could be a rock or a shipwreck.

Coral Reefs run in parallel with coasts.

They are generally found between the latitudes 30N and 30S because they can't survive in seas colder than 19 degrees Celsius (C).

Coral Polyps need oxygen and food, with the right conditions, to grow.

Conditions	Description
Temperatures	• Polyps grow best when the surface temperature averages between 22C and 25C • If sea temperatures rise due to global warming, existing coral reefs may die. However, new places will become habitable for polyps, meaning coral reefs may grow elsewhere
Water	• Polyps need clean, clear water to live in • Sediment in rivers prevent reefs from forming at river mouths • If the water is clouded with sediment, sunlight won't reach the plankton that the polyps feed on • Sediment settles on polyps – stopping them from feeding • Sediment is used in Corrasion during storms, damaging reefs • Corals grow best in high salinity – so won't grow as well in low salinity fresh water • Shipping and pollutants kill corals
Algae	• Coral can survive without sun, but they can't survive without algae • They need 1-cell algae to survive. This usually isn't found deeper than 10m, meaning reefs form in shallow waters

20C 30N-30S 10m Deep Clear Water

THEME 2
THE NATURAL ENVIRONMENT

Volcanic Island

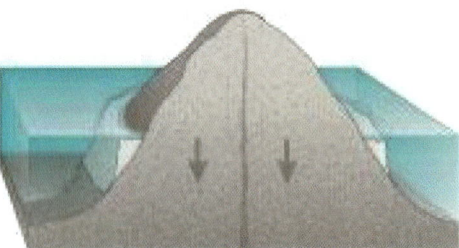
Fringing Reef

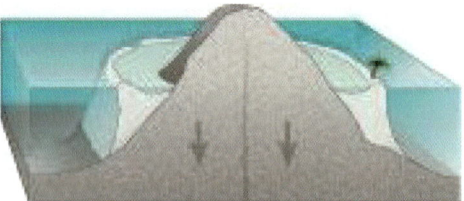
Barrier Reef

Atoll Reef

Figure 11

A fringing reef occurs when a previously volcanic island sinks down and a reef forms around the island, along with a shallow lagoon.

When the island sinks down further, the lagoon becomes wider and deeper, forming a barrier reef.

An atoll reef forms when the volcanic island has completely sunk beneath the surface of the water, leaving a shallow, flat floored lagoon, surrounded by reef, behind.

Fringing Reefs are the most common.

Barrier Reefs are the largest.

Atoll Reefs have no island.

2.3 COASTS

Mangrove Swamps

Describe … mangrove swamps and the conditions required for their development

Mangrove Swamps often form in the same areas as coral reefs because they have similar environmental requirements.

Characteristics

- **Stilt Roots** – these give the trees extra support in stormy weather. Also, they absorb extra oxygen from the air (because there is little in mud flats as the soil particles are closely packed)
- **Produce Flowers** – these pollinate, growing anchor roots
- **Drop Seed Capsules** – that carry seeds to other locations
- **Halophytic (salt loving)** – mangrove trees are specialised to survive in both salt and fresh water. This is because they are covered 2 times a day by saltwater
- **Dry at Low Tide, Flooded at High Tide**

They are good as:

- Nature Incubators
- Coastal Protection
- Wildlife Protectors

Image 28: Mangrove Forest, Puerto Rico

Conditions

They prefer to grow in sheltered places like river deltas, which have reduced wave and tidal action. Mangrove swamps grow on large, gently sloping mudflats.

24C+ in Warm Months

1250mm+ Rain Per Year   Calm / Sheltered Water

Conditions needed for Mangrove Swamps to form (above)

101

THEME 2
THE NATURAL ENVIRONMENT

Hazards and Opportunities

Demonstrate an understanding that coasts present hazards and offer opportunities for people

Hazards (including coastal erosion and tropical storms)

Hazards	Opportunities
• Coastal Erosion • Coastal Flooding – can destroy houses and infrastructure • Tropical Storms • Shipping hazards from hidden rocks • Tsunamis	• Recreation (beaches, sailing) • Tourism – generates money • Trade – ports • Wildlife Tourism – like in salt marshes and mangrove swamps • Using the sea for waste disposal • Fishing • Wave / Tidal Power

Tropical Storms

These can create a storm surge that floods large areas, destroying homes and infrastructure.

Tsunamis

These are created after an earthquake. Lowland areas near the coast are more vulnerable to tsunami damage.

Coastal Erosion

When there is rapid coastal erosion, land can collapse into the sea, destroying houses and infrastructure. This is more common where there are soft rocks, large waves and strong winds.

2.3 COASTS

Managing Impacts of Coastal Erosion
Explain what can be done to manage the impacts of coastal erosion

Some coastal areas have fragile ecosystems that need to be protected; others, factories or residential areas. Coastal management works to control erosion, flooding and longshore drift.

Hard Engineering techniques are often more expensive. They can damage the landscape and natural environment:

- **Sea Walls** – built along the base of cliffs to prevent erosion and flooding. They are curved to reflect waves. However, they can erode the beach away, they are expensive to build, are visually unsightly and carry a high cost of maintenance because the sea can erode the sea wall
- **Groynes** – barriers built at right angles to the beach to trap sediment and to prevent longshore drift. By maintaining beaches, you defend against erosion and preserve tourism. However, they are unsightly and expensive to maintain
- **Breakwaters** – large areas of boulders off the coast that reduce the power, or change the direction, of waves to reduce hydraulic action or longshore drift. The beaches and cliffs don't become unsightly. However, they can be unattractive and expensive to build
- **Gabions** – metal rock cages that absorb wave energy and are cheap. However, they have a short lifespan and are unsightly

An example of a breakwater

Soft Engineering techniques are usually cheaper and have a reduced impact on the landscape and natural environment:

- **Managed Retreat** – areas of coast are allowed to erode where there is low land value. The material that is eroded can then form beaches or salt marshes that reduce coastal erosion in the future. However, land is lost
- **Beach Nourishment** – replacing beach material that has been removed by erosion or longshore drift. Sand and shingle is transported to the beach, increasing its width and helping to preserve the beach in the future. However, the collection of the replacement sand used in this process could damage the environment.

THEME 2
THE NATURAL ENVIRONMENT

Case Study – New Forest Coastline, Hampshire, UK
The opportunities presented by an area or areas of coastline, the associated hazards and their management

Opportunities

- Tourism – people go to this area to see the beautiful beaches and to sail near the river estuaries
- Industry – because the area is on the coast, industry developed near Fawley due to good accessibility of trade links across the English Channel
- Housing – people love the views and incredible landscapes in this area
- Ecology – wildlife rich areas of scientific interest (like salt marshes) are useful for scientists studying salt marsh ecosystems

Hazards

- Cliffs are retreating at Highcliffe due to severe erosion. They are mainly composed of sand and clay, making them liable to erosion. In places, they have retreated by over 60 metres in the past 40 years
- Coastal erosion near Barton on Sea and Milford on Sea. Mudflows after heavy rain enhance this process. This area is built up. Land collapsing can destroy houses and infrastructure
- Hurst Spit can be eroded and breached during storms, destroying habitats
- Threat of flooding to the low lying coastal areas
- Coastal marshes are at risk from sewage and waste produced by major industries at Southampton Water
- Rubbish on beaches from tourism can damage habitats and kill wildlife

Management

- Building sea walls to protect the cliffs near Highcliffe. These break waves, widen the beach and protect the cliffs
- Building **dikes** (embankments) and flood walls to prevent severe flooding in the low-lying areas near Lymington
- Creating nature reserves to protect wildlife habitats from water pollution
- Encouraging sustainable tourism to reduce rubbish on beaches
- Building groynes to reduce the loss of beach from longshore drift. This also gives the cliffs extra protection, as they reflect some of the waves' energy

Image 29: Groynes at Hurst Castle on Hurst Spit

THEME 2: THE NATURAL ENVIRONMENT

WEATHER

Collecting Weather Data

Calculations

Graphs and Diagrams

THEME 2
THE NATURAL ENVIRONMENT

Describe how weather data is collected

Make calculations using information from weather instruments

Use and interpret graphs and other diagrams showing weather and climate data

So, what is Weather?

Weather is the state of the atmosphere at a certain time

Weather Elements are the parts of weather (e.g. Pressure)

Weather Events are a result of weather elements (e.g. Drought)

Elements of Weather and their Collection

Rain gauge, maximum-minimum thermometer, wet-and-dry bulb thermometer (hygrometer), sunshine recorder, barometer, anemometer and wind vane, along with simple digital instruments which can be used for weather observations; observations of types and amounts of cloud

All weather elements are linked, and all the energy that generates weather comes from the sun.

Sun -> Temperature -> Pressure -> Wind -> Humidity -> Cloud -> Precipitation

Image 30: Rain Gauge

Precipitation - Rain Gauge
Precipitation – Rain, Drizzle, Snow, Hail and Sleet

Measured with a **Rain Gauge**. This must be in an open space.

Recordings can be totalled to give a daily / monthly / annual rainfall.

Readings are read at the same time each day by pouring the collected water into a measuring cylinder.

They should be situated in an open space away from overhanging trees or vegetation that may drip into it, decreasing accuracy. The gauge is put into the ground for stability (or on a solid stand), with the top 30cm from the surface of the ground to reduce rain splash.

106

2.4 WEATHER

Temperature - Thermometers
Temperature – The degree of warmth in the air

Measured with a **Thermometer**.

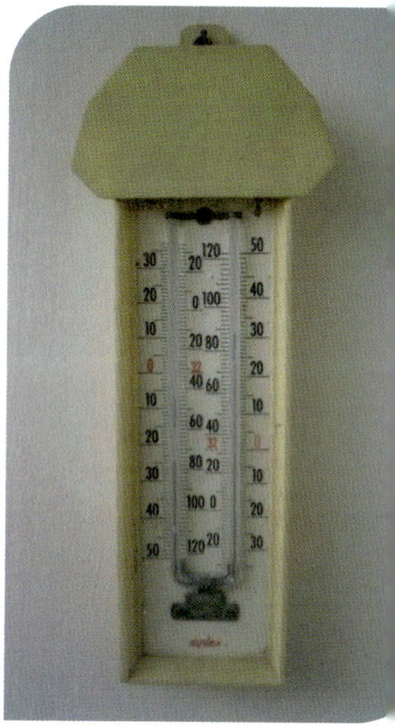

Image 31: Six's Thermometer

Maximum Thermometers use mercury – the level rises with heat

Minimum thermometers use alcohol – which contracts when temperatures fall

Six's thermometer combines the max. and min. thermometers

To measure air temperature, they should be shaded from sunlight and heat sources. They should also be raised and away from buildings to avoid residual heat affecting results.

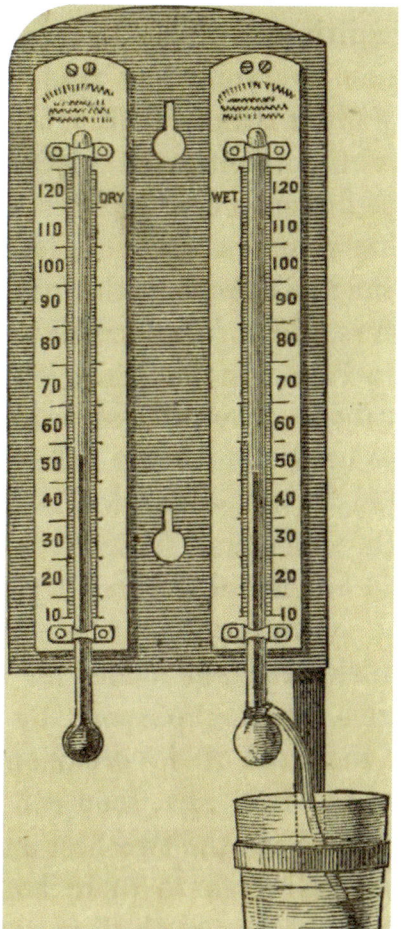

Image 32: Dry and Wet Bulb Thermometers

Humidity - Hygrometers
Humidity is the amount of water vapour in a given volume of air. Water vapour is invisible, and all air contains some quantity of it

The amount of water vapour air can hold is determined by the temperature.

Warm air can hold more water vapour than cold air.

1. When air temperature increases, water evaporates, increasing the air's humidity
2. When air temperature decreases past the dew point, condensation occurs

Knowing how close the air is to being **saturated** is crucial to predicting precipitation.

When the air is saturated, the relative humidity is 100%.

Relative Humidity (%) = (amount of vapour in air / level of water vapour needed for saturation) x 100

Relative humidity is measured using a **hygrometer** (a combination of a wet and dry bulb thermometer)

107

A dry bulb thermometer is a normal thermometer.

On a wet bulb thermometer, the bulb is wrapped in muslin cloth. A moist wick is fed into this from a water supply, meaning the bulb is surrounded by a moist muslin cloth.

> The Depression of the Wet Bulb = Temp. of Dry Bulb – Temp. of Wet Bulb

The larger the depression is, the lower the relative humidity is.

If the temperatures are equal in the wet and dry bulbs – the air is saturated.

You can use a **depression table** to work out the relative humidity from the wet and dry bulb temperatures.

Image 33: Sunshine Recorder, Botanical Garden, Funchal, Madeira

Sunshine - Sunshine Recorder
Sunshine hours are the number of hours in a day where there is sunlight.

A **sunshine recorder** records the number of hours of sunlight using a light sensor and a storage mechanism to record the number of hours of sun there were in a day.

They should be sited above the ground, out of the shade and (in the Northern Hemisphere) facing southwards so it is exposed to the sun through the day.

The sun's rays pass through a glass sphere and burn a piece of heat sensitive card. This is marked with hours and minutes that represent a day. As the sun moves across the sky, a line is burnt that shows when there was sunshine.

2.4 WEATHER

Pressure – Barometers

The atmosphere exerts **pressure** on the Earth's surface, as air has weight.

Pressure is the amount of weight exerted on the Earth's surface in a column of atmosphere.

Pressure is measured in Millibars (mb), and it varies from place to place and from time to time.

On average, the pressure at sea level is 1013 mb.

> If the pressure is greater than 1013 mb, it is called High Pressure
> If the pressure is lesser than 1013 mb, it is called Low Pressure

As the altitude increases, pressure decreases. This is because there is less air above the ground's surface at that point. So, recordings are converted into sea-level equivalents when plotted on isotope graphs. Otherwise, a pressure map would just be an inverted relief map, which wouldn't be as useful.

The pressure lines on a map are called **isobars**. They are drawn to connect areas with the same pressure – like contour lines, but for pressure instead of altitude.

There are **3** kinds of instruments used to measure pressure:

Mercury Barometer	Aneroid Barometer	Barograph
Pressure forces the mercury in the bowl downwards. This means the mercury is forced up this tube, giving a pressure reading	There is a corrugated box inside that shrinks and expands dependant on pressure. The pressure is shown using a dial on the outside of the box	An aneroid barometer with a revolving drum, paper and a pen. It is used to show pressure changes over time
Figure 12: Mercury Barometer	Image 34: Modern Aneroid Barometer	Image 35: Barograph

109

THEME 2
THE NATURAL ENVIRONMENT

Wind Speed - Anemometers

Wind is air moving over the surface of the ground.

Wind Speed is measured using an **anemometer**

This is sited 10m above the ground in an open space. The cups rotate around a shaft – which raises the anemometer above the ground, reducing ground air deflection.

The speed that the cups rotate when driven by wind is measured and shown on a meter in metres per second.

Wind speed changes, so you need to note any strong gusts to improve the accuracy of average measurements. Often, the data is recorded on a moving graph to show the **gusts** (fluctuations in wind speed) and to increase the reliability of the overall measurement.

Image 36: Anemometer

The **Beaufort Scale** sorts wind speeds into classes, categorised by the damage caused.

Image 37: Wind Vane, Farnborough

Wind Direction - Wind Vanes

Wind blows in different directions.

The **Prevailing Wind** is the most common wind direction in an area.

A **Wind Vane** has points of the compass fixed and accurately sited to point in the correct direction. Above the compass sits an arrow that can move in light winds.

When the wind moves, the thickened end is pushed so the arrow points to where the wind came from.

The direction of the wind helps meteorologists to work out how the weather will change.

110

2.4 WEATHER

A wind vane should be sited more than 10 metres above the ground in an open space (3 times further away from the nearest object than the nearest object's height).

Ideally, they should be sited on top of buildings. This is so the arrow can move freely in all directions without interference.

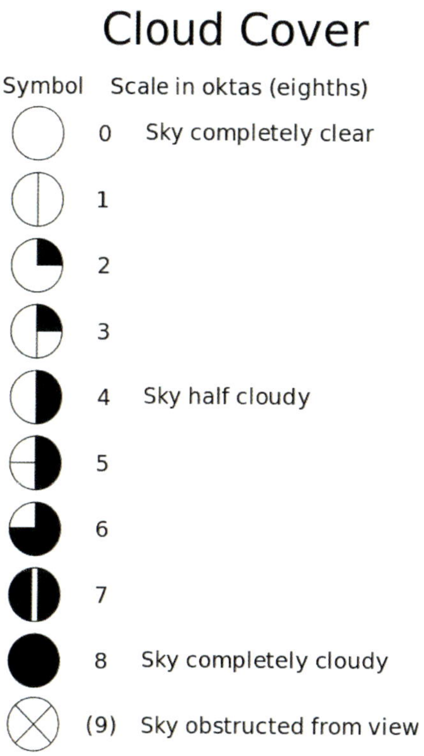

Clouds

The amount type of cloud present is recorded by **observation**.

A code of **symbols** is needed to compare clouds across different regions or countries.

Cloud cover is estimated by looking up at the sky, dividing it into eighths (oktas) and judging how many of the eighths are covered by clouds.

There are 3 main types of clouds:

- **Cirrus** – High altitude 'wispy' clouds
- **Stratus** – Like a sheet
- **Cumulus** – Rippled shape with rounded tops

Cloud Cover

Symbol	Scale in oktas (eighths)
○	0 Sky completely clear
◐	1
◔	2
◕	3
◑	4 Sky half cloudy
◕	5
◕	6
◕	7
●	8 Sky completely cloudy
⊗	(9) Sky obstructed from view

Figure 13: Cloud Cover Symbols

Figure 14: Cloud Types

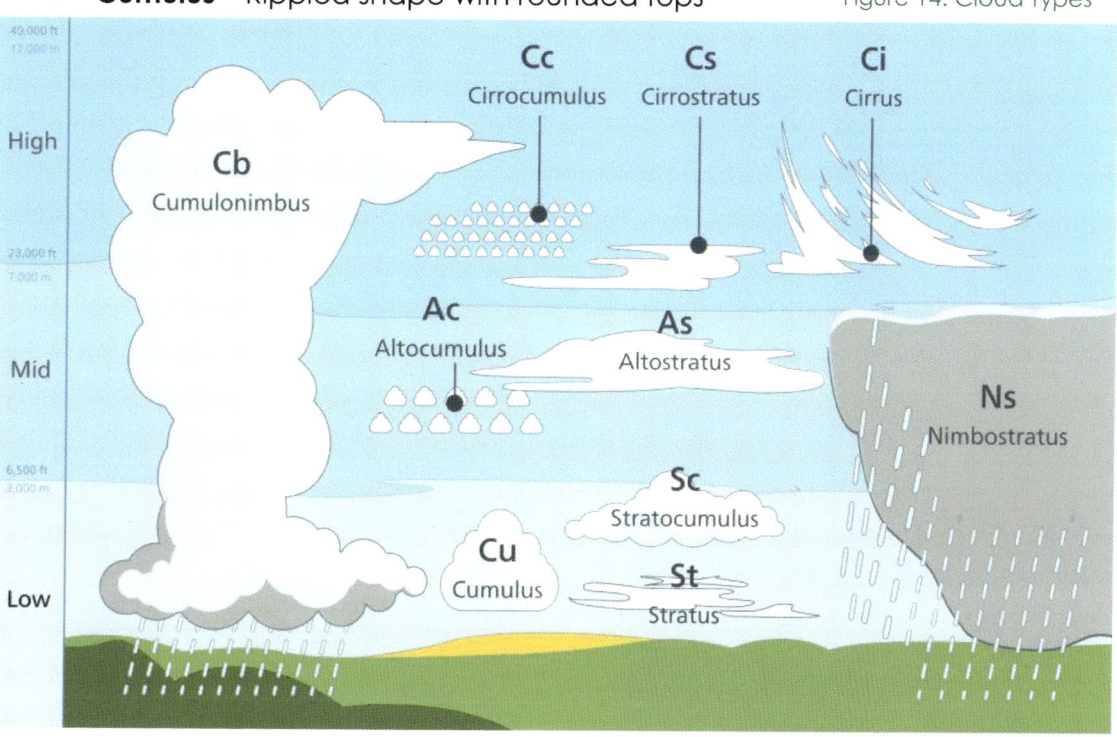

THEME 2
THE NATURAL ENVIRONMENT

Clouds form when **water vapour** condenses into water droplets in the atmosphere. This happens when the air is saturated because the air can't hold any more water.

Digital Instruments

Nowadays, weather data is more frequently being collected automatically by computers.

This increases reliability because computers won't make errors when reading weather data.

It also means data can be collected more frequently – as computers can collect the data automatically and more quickly than a meteorologist, who has to walk over to the instrument to manually collect the data.

Also, the data can be collected remotely, meaning weather stations can be sited in more remote collections. This allows for better weather observations and consequently, better weather prediction.

In addition, digital data can be analysed deeply by supercomputers to better predict and extrapolate weather conditions to produce even better predictions. This benefits shipping and air traffic control.

Stevenson Screens

Describe and explain the characteristics, siting and use made of a Stevenson screen

Maximum-minimum thermometers and hygrometers are housed in a special wooden box called a **Stevenson screen**. Barometers are often put inside it to protect them from the elements.

This box is designed to ensure the instruments give the correct readings.

Air temperatures need to be measured in the shade – out of direct sunlight – and at a standard height above the ground. This is so temperature data to be compared, and so instruments aren't affected from heat radiating from the ground.

The screen should be positioned so that the access hatch opens away from the sun (facing the direction at the start of the hemisphere name, North in the Northern Hemisphere).

There are also slatted sides called **louvres** that permit the free flow of air in and out of the screen. This is needed to measure the outside air temperature, and to allow evaporation from the hygrometer to correctly measure the air's humidity.

Features:

- Slanted roof sheds rain
- Double layer of wood on the roof increases durability from

2.4 WEATHER

the rain, and so the air space stays at a constant temperature
- Louvres permit free air movement
- The airspace provides insulation
- The box is white to reflect the sun's rays
- It is also raised on legs to avoid heat radiated from the ground and to ensure the thermometer bulbs are at the standard height of **125cm**
- Built of wood, so heat isn't conducted inside the screen.

NB Grass reflects (and thus, radiates) less heat than concrete surfaces.

Image 38: Stevenson Screen

Calculations + Interpreting Graphs and Diagrams

Make calculations using information from weather instruments

Use and interpret graphs and other diagrams showing weather and climate data

After weather data is collected, it is manipulated by making calculations of useful or relevant statistics, or by presenting it in graphs and diagrams to allow meteorologists to more easily identify patterns in the data.

In an exam, you could be asked to manipulate weather data, climate data, draw graphs or even use a relative humidity table to work out humidity.

If you are asked to **compare** data, compare it! Remember to use data in your answer. Ensure you know what the data specific **command words** mean, so you fully understand what the question is asking about.

This section involves calculations. To revise this section, I recommend checking out the past papers on the CIE website at http://www.cie.org.uk/programmes-and-qualifications/cambridge-igcse-geography-0460/past-papers/ or asking your teacher for past exam questions that involve processing and analysing weather data.

THEME 2
THE NATURAL ENVIRONMENT

THEME 2: THE NATURAL ENVIRONMENT

CLIMATE

Equatorial

Hot Desert

Characteristics

Reasons for locations of climates

Climate Graphs

Ecosystems

Deforestation

Case Studies

THEME 2
THE NATURAL ENVIRONMENT

Describe and explain the characteristics of two climates: • equatorial • hot desert

Describe and explain the characteristics of tropical rainforest and hot desert ecosystems

Describe the causes and effects of deforestation of tropical rainforest

Weather is the short term day-to-day changes in the atmosphere for a particular place. This could be rainfall or temperature.

Climate is the average weather conditions over a period (e.g. over a few decades). Climate regions are over larger areas, not just a specific place.

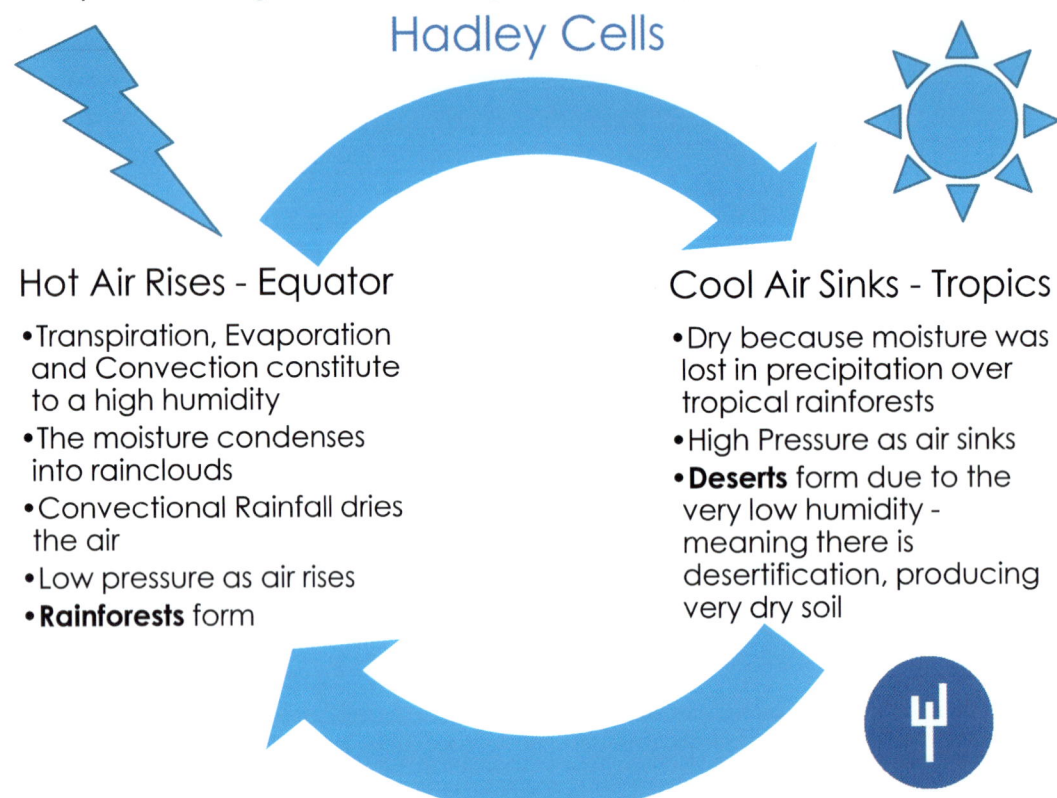

Hadley Cells

Air moves in cells in the Earth's atmosphere. These atmospheric cells create conditions that form deserts at the tropics and rainforests at the equator.

The **2** cells that straddle the Equator – between the Equator and each tropic – are called **Hadley cells**.

116

2.5 CLIMATE

These cells are responsible for the different climates.

NB   Diurnal temperature ranges

are the ranges of temperature during a day (i.e. differences between Night and Day temperatures)

Characteristics

Climate characteristics (including temperature [mean temperature of the hottest month, mean temperature of the coolest month, annual range]; and precipitation [the amount and seasonal distribution])

Characteristic	Equatorial	Hot Desert
Mean Temp (Hottest Month)	25-30C	30-40C
Mean Temp (Coolest Month)	25-30C	10-20C
Annual Temp Range	Constant (around 3C)	Extreme in day, constant through year
Precipitation / Year	Over 2000mm	Very low
Distribution of Precipitation	High rainfall throughout year. Sometimes there is a wet / dry season	Very low, especially in summer

Equatorial Climates

Tropical Rainforests are found between the tropics (28C N/S), near the equator

The main concentrations are in:

- The Amazon Basin
- South East Asia
- Madagascar
- S. India + Sri Lanka
- Central America
- The Congo Basin
- NE Australia

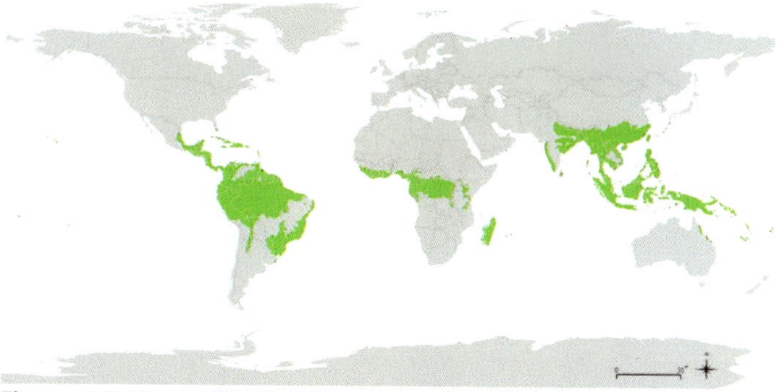

Figure 15: World Distribution of Tropical Rainforests

Characteristics:

- High annual temperatures throughout the year
- Frequent heavy rainfall
- Convectional rainfall
- A greater diurnal temperature range than annual temperature range
- No obvious seasons
- High humidity

THEME 2
THE NATURAL ENVIRONMENT

Factors affecting characteristics
Factors influencing the characteristics of these climates (including latitude, pressure systems, winds, distance from the sea, altitude and ocean currents)

The **Hadley cells** straddle the Earth's equator. Moist air rises at the equator, condensing into clouds. When the air rises, it creates low pressure – so, tropical rainforests are in areas of low pressure. Moist air rises because hot temperatures warm up the land during the mornings: causing evaporation, transpiration and convection. This produces the typical convectional rainfall in the afternoons.

Days start off warm, with the temperature and humidity increasing throughout the day (due to transpiration, evaporation and convection). This makes the air saturated by the afternoon, forming heavy rainstorms.

The air is now much drier (because of the heavy rainstorms decreasing humidity), sinking back down at the tropics. This forms desert environments there because the air is very dry.

The world's rainforests are located between the tropic of cancer (N) and the tropic of Capricorn (S). As they are near the equator, the sun is high in the sky throughout the year. This means that there is very intense heat and light from the sun through the year, producing annual high temperatures in equatorial climates.

Rainforests are often close to the coast, or near rivers, as air currents and groundwater bring moisture into the environment. Combined with the sun, hot and humid environments ideal for tropical rainforest environment are formed. Wind can also contribute to the formation of a tropical climate, as it can blow warm, moist air from hot seas to rainforests.

Image 39: Rainforest in Belize

At a lower altitude, air can become warmer than if it were at a higher altitude, this is because air is heated by the ground – so air closer to the ground is heated more than air further from the ground. Rainforests usually grow in areas of low altitude because the air can become warmer.

2.5 CLIMATE

So, tropical rainforests are generally found between the tropics because this is where sunlight, warmth and rainfall (the ingredients for life) are at their greatest.

Hot Desert Climates

Hot Deserts are found at the tropics (28C N/S).

The main concentrations are in:

- North Africa
- W. North America
- W. South America
- W. South Africa
- Central Australia
- Middle East
- Central Asia

Characteristics:

- Absence of cloud cover – no rain, high diurnal temperature ranges
- High pressure – meaning that clouds can't form
- Arid due to the climate and sparse rain

Image 40: Dunes in The Sahara Desert

THEME 2
THE NATURAL ENVIRONMENT

Factors affecting characteristics
Factors influencing the characteristics of these climates (including latitude, pressure systems, winds, distance from the sea, altitude and ocean currents)

The **Hadley cells** straddle the Earth's equator. Moist air rises at the equator, but the moisture is then lost through precipitation. This air proceeds to cool and sink over the tropics. As this air is sinking, high pressure is created at the tropics. This high pressure means that there is no condensation – so there are no clouds. This also means that there is little or no evaporation or transpiration. As a result, there is often little or no surface water, so deserts are formed.

Also, air heats as it descends – making the tropics even hotter and dryer, meaning deserts are extremely **arid** (dry).

The lack of condensation means that there is no cloud cover – so days are hot (there is no shade) and nights cold (heat is lost due to a lack of clouds). This means that deserts have high diurnal temperature ranges.

The latitude also affects deserts. The sun is never low during the year because deserts are at the tropics. This means that winters are much milder than at higher or lower latitudes (like in Europe or Southern South America). However, there is some seasonal variation, because the sun's position is slightly lower in the winter and slightly higher in the summer. It is therefore hot in hot tropical deserts through the year.

Deserts are often far from the sea. This is because nearer the sea air can blow onto the land – bringing in moisture. This means that the further you are from the sea, the drier it becomes. Therefore, many deserts are in land-locked countries where there is no moisture that can come onto the land from the sea. Also, ocean currents at the tropics can be cold – so the sea water is less likely to evaporate and transfer moisture to the land.

Climatic Graphs
Climatic graphs showing the main characteristics of temperature and rainfall of the two climates

[Manaus – located in the Amazon Rainforest – see next page] Notice that the annual temperature range is under 1.5C, so there are constant temperatures throughout the year in Equatorial Climates. Also, note that there is high precipitation throughout the year, but with a wet season in March and a dry season in August. This is normal for an Equatorial Climate.

2.5 CLIMATE

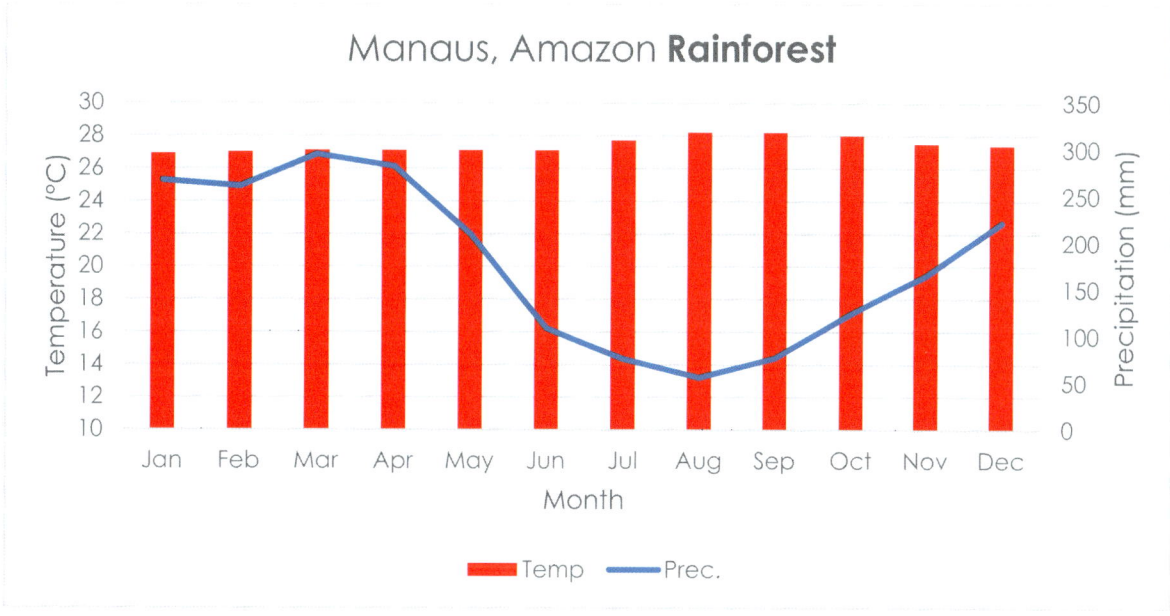

Climate Graph of the typical climate of Manaus, Amazon Rainforest

[Tazirbu – in the Sahara Desert – see below] Notice that the temperature is greater in the summer, and that there is seasonal variation, of about 18C, because the sun is lower in the winter at the tropics (making them cooler). Also, note that there is very little precipitation throughout the year, creating an arid environment.

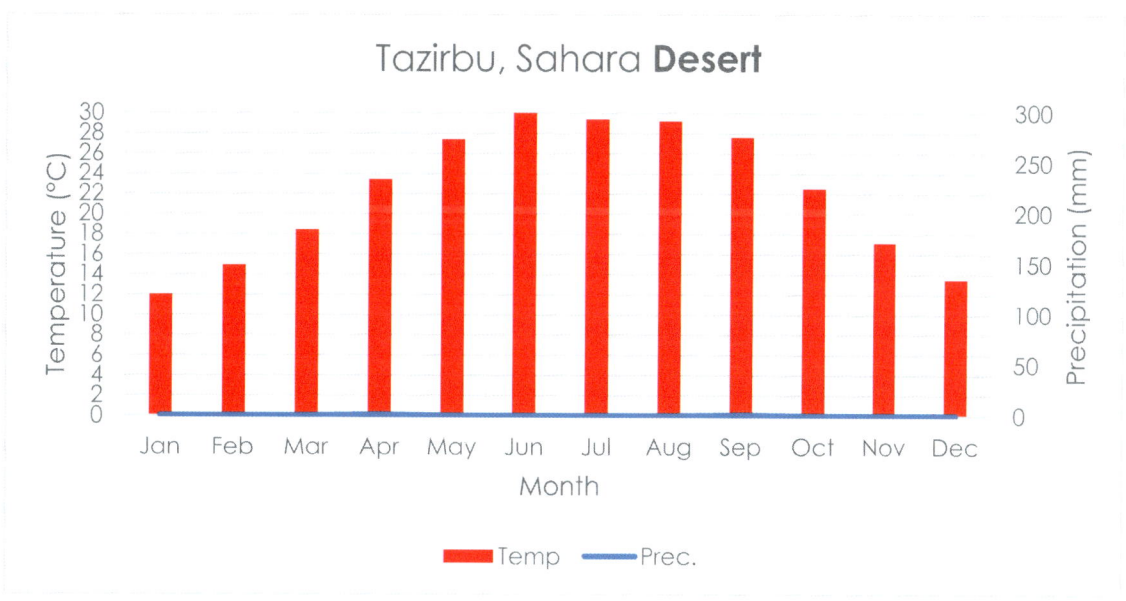

Climate Graph of the typical climate of Tazirbu, Sahara Desert

121

THEME 2
THE NATURAL ENVIRONMENT

Ecosystems

Describe and explain the characteristics of tropical rainforest and hot desert ecosystems

The relationship in each ecosystem of natural vegetation, soil, wildlife and climate

An **ecosystem** is a complex system of plants and animals that depend on each other and the environment.

A **biome** is a large ecosystem. **Biospheres** are made up of components (living or dead) that depend and interact with one another. As they depend on the environment, ecosystems are specific to different places.

Tropical Rainforest

The tropical rainforest biome has the ideal conditions for life, which leads to the creation of an advanced and rather complex ecosystem. The conditions are formed by heavy rainfall (plentiful water) and the hot equatorial climate at a stable temperature. This means that rainforest biomes contain a higher proportion of organisms and life than anywhere else on Earth.

All the plants compete for light – because plants need to photosynthesise. This fight means that rainforest vegetation grows in layers. Most plants and animals live in the canopy because there is not much available light beneath the canopy.

Image 41: Rainforest at Puentes Colgantes, near Arenal Volcano, Costa Rica

2.5 CLIMATE

Characteristics

Structure

Emergents: Very tall trees that grow above the canopy layer to reach sunlight

Canopy: The top layer of the rainforest, containing around half of its wildlife. This layer is extremely thick and is often over 30m deep

Shrub Layer: This is the layer nearest the canopy floor. It is very dense (as it collects dead material falling from the canopy). The organisms here have to adapt to live in less sunlight, because the canopy blocks light

Humus Layer: Made up of decaying organic matter. As rainforest soil is infertile, the rainforest depends on the humus layer for its nutrients

NB   Humus Layer

Humus is made up of **decaying organic matter**. Without this layer, the soil structure (and nutrients) would be lost, inducing soil erosion and loss of wildlife.

Plant adaptations:

- **Buttress roots** – these are very long roots that come off tall trees. These are meant to capture the most nutrients as possible from the poor rainforest soil. Most smaller, non-rainforest trees have shallow roots which would be unable to support the largest rainforest trees. Buttresses help to hold the heavy trees up by providing additional support.
- **Drip-tip leaves** – some plants develop pointed tips to their glossy leaves to remove water from them more effectively (this is because rainfall is so frequent and heavy)
- **Lianas** – woody vines which climb up trees to reach sunlight
- **Strangler Figs** – these wrap around a host tree, taking their light and nutrients
- **Evergreen state** – as rainforest have stable climates, trees can shed their leaves at any time of year. Constant death and rebirth of different trees at different times creates organic matter – composed of dead leaves. This is why rainforests look so green throughout the year.

THEME 2
THE NATURAL ENVIRONMENT

Nutrient Cycle

Producers – plants capture the sun's energy and store it as glucose or starch to make the energy available to consumers in the rainforest.

Consumers – these feed on the producers and any consumer in a lower trophic level (lower down) in the food web.

As nutrients are heavily sought after by the rainforest's plants, nutrients are sparse. This means that nutrients must be recycled, the survival of the rainforest is very much dependent on it:

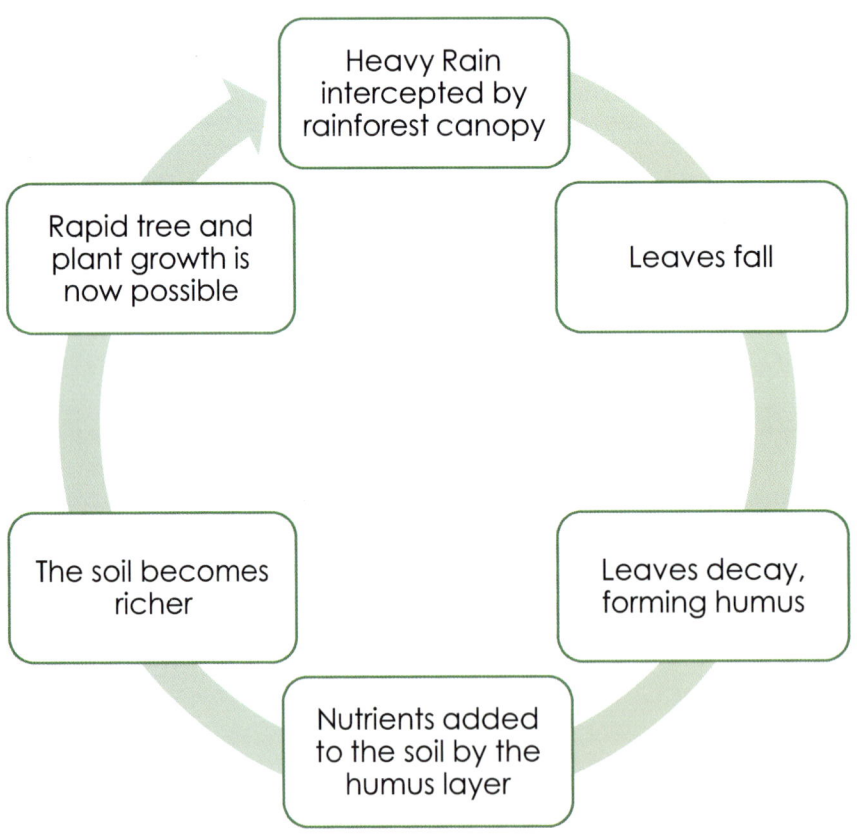

2.5 CLIMATE

Hot Desert

Deserts, as we learnt earlier, are hazardous and arid environments which are hostile to life.

In some deserts (e.g. Yuma Desert) there is little life because there is no water or available nutrients.

However, some deserts can support life. This would depend on the:

- Frequency and amount of rainfall
- Amount of nutrients in the soil
- Presence of groundwater

Even if there is rainfall, there will be drought periods. Plants, like grasses, lie dormant during a drought. The seeds lie inactive until the next rainfall, and rapidly develop when they receive water to make the most of an opportunity for growth.

If there is water (like in an **Oasis**, which form when the water able is near the surface), plants can grow. Oases are often full of tall, dense vegetation supported by the constant sun, available water and high temperatures.

However, in most cases, there is only light, rare and unreliable rainfall. Because the soils are baked and hardened by the sun, there will be little water infiltration – forming a low water table. There will be high surface runoff, so little water will be stored, meaning little water will actually be available to plants when it eventually rains. Due to the intense sun, there is high evaporation, meaning plants will need more water than usual to survive. Desert soils are also thin, meaning that dense vegetation can't be supported. As the vegetation is consequently sparse, there is little organic material available to form humus. The little soil there is also eroded by harsh desert winds. The large diurnal temperature ranges

Image 42: Arizona Desert

also mean that plants will have to be specially adapted to live in these hostile environments.

Plants can do this in a few ways:

- **Water retention** – some plants, like cacti can store water in their trunks in wet weather, making it available for use in a period of drought
- **Leaf shape** – many desert plants have adapted to reduce their water loss, so they can store more water and be shaded from intense sunlight. Cacti spines have a small surface area. reducing transpiration
- **Water collection methods** – some plants have developed unique methods of collecting water. Some plants have multiple inlets which independently collect water for the plant to use, making the most of the sparse water
- **Tolerance of salty soils** – salt is normally toxic to plants. Plants in deserts adapt to cope with the desert's high salinity from high evaporation
- **Flowering times** – some plants only flower and transpire at night. This is to reduce the impact of the harsh sun, meaning water can be used more effectively. This allows some plants to survive with less water.

Cacti (a plant commonly found in deserts globally):

- Fine Hairs – reduces transpiration and water loss
- Water Storage – means they can better survive in times of drought
- Non-Classical Leaf Shape – reduces water loss

Camels (an example of animal adaptations):

- Long Eyelashes to keep sand out of eyes
- Can close nostrils for protection in sandstorms
- Can go without water for months
- Can drink lots of water quickly when it is available
- Store fat in humps because food isn't readily available in deserts
- When fat is digested, hydrogen is released. This mixes with oxygen from the air to form water

Image 43: Camel in Desert

2.5 CLIMATE

Deforestation of Tropical Rainforest

Describe the causes and effects of deforestation of tropical rainforest

Effects on the natural environment (both locally and globally) along with effects on people

Around **5%** of the Earth's surface is covered in Tropical Rainforests **today** compared to around **15% 50 years ago**

Deforestation – removal or cutting of trees

Reforestation – replacement of trees that have been cut down

Afforestation – planting of trees in a location where there were no trees before

Causes
- **Cattle Ranching**
    - Lots of rainforest has been burnt down for cattle farming (**slash and burn**), to supply cheap beef to countries like America and China
    - This slash and burn farming accounts for about half of deforestation
    - However, after the rainforest trees have been removed, the land can't be used for long because the soil quality is so poor. The soil quickly becomes dry, with heavy soil erosion, leading to desertification.
    - The cattle farmers then have to move on to new rainforest, and do more slashing and burning to create new cattle pastures
- **Logging** (often for Hardwood)
    - The 2nd biggest cause of deforestation
    - Timber companies cut down massive trees which are valuable in MEDCs – like Mahogany – to make furniture
    - Logging companies often use **Clear felling** (cutting large areas at once). Only the most valuable trees (e.g. mahogany) are kept
    - Roads are also built to transport the timber, inflicting further damage on the rainforest
- **Agriculture**
    - Many fruits and cereals that we buy from tropical countries are grown in deforested areas
    - Rainforest is cut down to make room for massive plantations
    - Like with Cattle Ranching, this soil will quickly run out of nutrients (rainforest soil is naturally poor), and will no longer be able to sustain crops
    - So, farmers go on to deforest more rainforest for more food plantations

- **Mining**
    - MEDCs need minerals and metals for high technology industries
    - These are often found in the ground beneath rainforests
    - So, the rainforest must be removed to extract the minerals
    - Chemicals used in extraction (like mercury) can enter water supplies, killing wildlife
- **Dams**
    - Some dam projects use large rainforest rivers (e.g. Amazon) for HEP power production
    - This could lead to massive flooding of rainforest in the creation of a reservoir
    - Dams could become blocked with soil washed from deforested land, producing flooding and further damage

Effects

The destruction of rainforests has significant impacts on the environment.

As the rainforest soaks up significant amounts of our CO_2 emissions, the loss of the rainforest will mean that less CO_2 is being removed from the atmosphere. Additionally, the burning of a rainforest in a 'Slash and Burn' manner will only add more CO_2 into the atmosphere.

This extra CO_2 could enhance the global warming effect. The warming of the atmosphere would lead to the melting of ice caps, sea level rise (from warmer seas) and the destruction of coastal habitats (because of increased coastal flooding from the increased sea level), including coral reefs by coral bleaching.

Sea levels may rise by up to 40 cm by 2100. The rising temperatures could also make species, that are unable to adapt quickly enough, extinct. Enhanced global warming may also lead to an increase in tropical storms, hurricanes and drought.

Burning Trees + Reduction of CO_2 Absorption -> More CO_2 in the Atmosphere -> Enhanced Global Warming -> Melting Ice Caps + Sea Level Rise -> Coastal Flooding -> Loss of Habitat -> Potential Extinction -> Food Chains are Impacted

Why Protect the Rainforests?
- They help with flood control
- They help to prevent soil erosion / providing soil structure
- They may contain medical remedies
- Ecotourism – a source of revenue in tropical countries
- Biodiversity – rainforests cover 5% of the Earth's surface, containing half of the world's biodiversity

2.5 CLIMATE

How?
- Zero Deforestation Policies – using sustainable palm oil, paper, food (e.g. beef) and timber sources that don't lead to deforestation
- Protecting the right to land for indigenous people
- Promoting sustainable choices to the public

Image 44: Charcoal Production and Deforestation

Case Study – Tropical Rainforest – Bornean Rainforest
An area of tropical rainforest

Borneo is located on the Equator, meaning that it has an equatorial climate. Consequently, most of Borneo is covered by tropical rainforest vegetation.

The island has a low population density and is the world's largest grower and exporter of palm oil. This is commonly used in cooking and cosmetics.

Borneo has 2 characteristics that lead to it having an Equatorial climate:

1. High temperatures throughout the year, with a low annual temperature range
2. Heavy rainfall throughout the year

This means that the air is very humid.

However, most of the rain falls during the monsoon season, which is common in Asia (in Borneo, it lasts from Nov-Jan). This is when moist air is blown in from the Pacific Ocean (NE of Borneo) due to changing wind systems. Also, the dry months are around April, this is because of air being blown from Australian Deserts (SE). Thus, the wind direction affects the humidity in Borneo, consequently affecting annual rainfall patterns.

THEME 2
THE NATURAL ENVIRONMENT

The Bornean Rainforest is also special because of its wildlife and unique ecosystem. It is the natural habitat for Bornean Orangutans. However, their population has halved in the last half-century. This is because their habitat is being destroyed by deforestation, they are being poached for illegal petting and are sometimes killed when farmers move to protect their crops.

Additionally, because of large scale logging since the 1970s, the rainforest has been reduced even further. There are 3 main reasons for this:

1. Destruction of rainforest to provide land for growing palm oil
2. Destruction of rainforest (in slash and burn) to provide extra farmland to feed the growing Malaysian population
3. Demand for wood in Malaysia's (and the world's) paper pulp, hardwood and plywood industries

National Parks have now been created to protect some of the rainforest. Additionally, the Heart of Borneo Conservation Plan (organised by the Malaysian, Bruneian and Indonesian governments) aims to help protect over 200,000 km² of forest habitat.

Image 45: Male Bornean Orangutan

2.5 CLIMATE

Case Study – Hot Desert – Namib Desert, Namibia
An area of hot desert

This desert is located near the West coast of Namibia.

In the south of the Namib desert, there are some of the tallest sand dunes in the world – however, in the north, the desert is bare and gravelly. As in most deserts, there isn't any surface water apart from after rare, unpredictable rainfalls.

The climate is coolest along the coast because of cool prevailing winds blowing off the Atlantic Ocean. These winds are cold due to the cold currents in the Atlantic Ocean (West of the desert), like the Benguela Current.

Warmer air in the desert forms as a result of the Drakensberg Mountains' rain shadow effect. As the warm, moist air from the Pacific Ocean (East of the desert) crosses the Drakensberg Mountain Range, it rises. As it does so, the air cools with the moisture condensing into clouds. The dryer air now warms and descends westwards down to the Namib desert. This is the **rain shadow effect**. The Namib lies in the 'shadow' of the Drakensberg Mountains, leading to little rain (due a low humidity) and high temperatures (from the sinking, prevailing oceanic air).

When the warm dry air from the desert meets the moist, cold air at the coast, coastal fogs and temperature variations are formed. The climate is cold near the Namib's coast, due to the Benguela Current, and hot elsewhere in the desert (because of the hot, dry air descending from the Drakensberg Mountain range). Consequently, as there are few days with rainfall and the air is warm, the climate is arid.

[Continued on the next page]

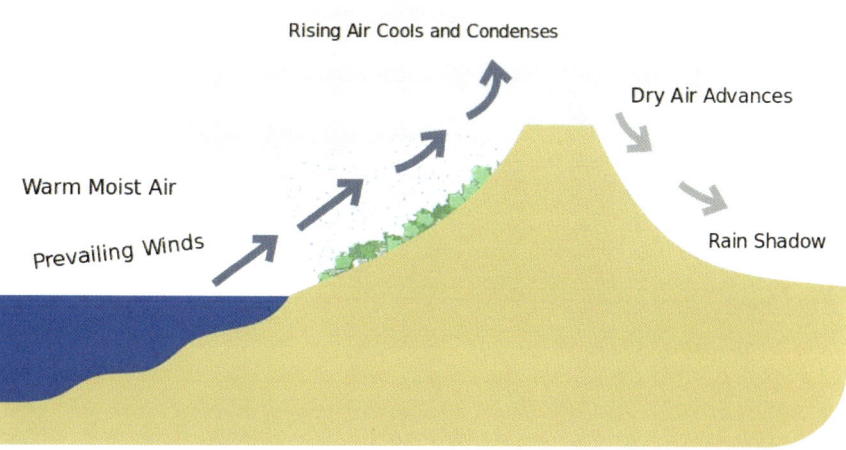

Figure 16: Rain shadow effect

THEME 2
THE NATURAL ENVIRONMENT

The most important feature of the Namib's climate is the famous coastal fog. This is formed when the Benguela current's cold air interacts with both the warm air from the Hadley Cell and the warm air in the mountains' rain shadow. This is the source of water for much of the Namib's wildlife.

The Welwitschia shrub plant is well adapted to the Namib's extremely arid conditions. It gets its moisture from the daily sea fogs. The other plants are adapted to deserts as you would usually expect, with small surface areas and high water retention. However, it also collects groundwater.

The Namib Desert beetle's body encourages the morning fogs to condense into water droplets onto its shell. The droplets roll down their backs to their mouths. There are some other beetles that gather water in a similar way in the Namib – they are called 'fog beetles'. This adaptation allows the beetles to survive in such an arid environment, the fogs provide them with a reliable water source.

2.5 CLIMATE

Image 46: Dunes in the Namib Desert

ECONOMIC DEVELOPMENT

THEME 3: ECONOMIC DEVELOPMENT

THEME 3
ECONOMIC DEVELOPMENT

- Use a variety of indicators to assess the level of development of a country
- Identify and explain inequalities between and within countries
- Classify production into different sectors and give illustrations of each
- Describe and explain how the proportions employed in each sector vary according to the level of development
- Describe and explain the process of globalisation, and consider its impacts

Development Indicators

- Use a variety of indicators to assess the level of development of a country
- Indicators of development (including GNP per capita, literacy, life expectancy and composite indices, e.g. Human Development Index (HDI))

Indicators of Development

Indicator	Conclusions
Birth Rate	Low – More Developed – better healthcare services High – Less Developed – traditional families – lots of children
Energy Consumption	Low – Less Developed – less resources High – More Developed – can afford luxuries, need more power, technological availability
Number of People per Doctor	Low – More Developed – better healthcare services High – Less Developed – less doctors due to lower literacy
Urban Population	Low – Perhaps less developed High – Perhaps more developed
Life Expectancy	Low – Less Developed – poor healthcare services High – More Developed – good healthcare services
Death Rate	Low – More Developed – better sanitation High – Less Developed – worse sanitation
Infant Mortality	Low – More Developed – good healthcare services High – Less Developed – poor healthcare services
Adult Literacy Rate	Low – Less Developed - people might not be able to afford education High – More Developed – people might be able to afford education

GNP (Gross National Product) per Capita is a measure of the wealth or money generated by a country. A country with a higher GNP is typically more developed than a country with a lower GNP.

3.1 DEVELOPMENT

Composite Indices

The **Human Development Index (HDI)** blends together 3 of the more important development indicators [Adult Literacy, Life Expectancy and GDP per Capita]. It was designed by the United Nations to provide a uniform system of determining the level of a country's development. Because it is standardised, it allows analysts to compare the development of different countries. The scale runs from 0 (least developed) to 1 (most developed).

The **Quality of Life Index** uses GDP, Life Expectancy, Happiness and how people feel about their family, workplace and local community. A country with a higher value is happier, and often more developed, than a country with a lower value.

Inequalities

Identify and explain inequalities between and within countries

Why is there inequality?

Income   Diet   Communications   Exploitation   Transportation
Gender   Education   Water Supply   Resources   Debt
Ethnicity   Religion   Energy Supply   Land Ownership
Corruption   Employment   Globalisation

Factors that constitute to inequalities between countries:

Physical	Human
<ul><li>**Climate** – Drought, Disease</li><li>**Natural Hazards** – Flood, Drought, Tsunami, Hurricanes</li><li>**Natural Resources** – Minerals, Natural Beauty (Tourism)</li><li>**Whether the Country is Landlocked** – Restricts Trade</li></ul>	<ul><li>**Colonialism** – Cheap resources, damage to culture</li><li>**Corruption** – Elite groups get given money by the government. The Military controls development</li><li>**Unfair World Trade** – LEDCs compete for the lowest price. MEDCs earn money from processing goods, so MEDCs get more foreign investment</li><li>**Poor Countries are in Debt to Rich Countries**</li><li>**Poor Countries can't invest** – like in Healthcare or Education</li></ul>

THEME 3
ECONOMIC DEVELOPMENT

Industrial Sectors

Classify production into different sectors and give illustrations of each

Primary, secondary, tertiary and quaternary sectors

Describe and explain how the proportions employed in each sector vary according to the level of development

Use of indicators of development and employment structure to compare countries at different levels of economic development and over time

Primary Sector	Secondary Sector	Tertiary Sector	Quarternary Sector
• **Collecting Raw Materials** • Farming • Mining	• **Manufacturing** • Cars • Computers • Phones	• **Providing a service** • Banking • Teaching • Lawyers	• **Knowledge Based Jobs** • Research • High Tech Industries (like Google)

LEDCs (Less Economically Developed Countries) have more people working in primary sector jobs. Meanwhile, MEDCs (More Economically Developed Countries) have more people working in tertiary sector jobs.

So, as a country develops, more people will start to work in a higher sector tier. This also means that the higher the HDI of a country, the more likely it is for people to be working in higher sector industries (like Education) than in lower sector industries (like Mining). This could be because of technology replacing Primary and Secondary sector jobs, which previously employed people. These people consequently have to work in the Tertiary sector.

LEDCs – More work in the Primary sector (farming is more important because people farm for themselves). More of the work is done by hand than by machines – perhaps because machines aren't readily available

NICs – More work in the Secondary sector (industrial development means people can manufacture products) and the Tertiary Sector (because people need to provide support services for the products manufactures in Secondary Sector industries)

MEDCs – More work in the Tertiary sector as machines have replaced jobs in the Primary and Secondary sectors with many goods now imported.

3.1 DEVELOPMENT

The employment structure changes over time throughout the development of a country. As a country develops, more people work in the secondary factor, migrating to urban areas to work in factories and industries. Mechanisation, becoming more common as a country develops, means that less people need to work on the land or in factories. People can also then afford to buy goods from other countries – further reducing the number of people employed in primary and secondary sectors.

As services develop, there is an increase in the number of jobs in public services and in tourism.

So, as a country becomes an MEDC, most of the workers are employed in tertiary industries. This could be because people would prefer an enjoyable, dynamic, well-paid job rather than a job involving physical labour.

As a country becomes even more developed, quaternary industries develop – this is because industries want to carry out Research and Development (R&D) to develop their products and services.

In the exam, you could be asked to study graphs or diagrams comparing employment structures of different countries. It is important to link the changes in amounts of people working in each industrial sector with the process of development.

Image 47: FCAB Freight Train, Bolivia

Globalisation

Describe and explain the process of globalisation, and consider its impacts

The role of technology and transnational corporations in globalisation along with economic factors which give rise to globalisation

Globalisation is the process by which the world is becoming more interconnected due to increased trade, with more cultural and economic exchanges.

The process of Globalisation

Where products are located all over the world and are transported around the world.

Globalisation is principally aided by **transportation** (e.g. Ships and Planes) and **technology** (e.g. the Internet and Computers).

Advances in technology ease the communication that is needed to exchange goods. If people can better communicate on a global

139

THEME 3
ECONOMIC DEVELOPMENT

scale, they can more easily control production and distribution all over the world. The internet is a key facilitator of this, allowing centralised control of global operations.

This permits increased connections across the globe that could lead to the increased globalisation of companies, services and goods. These connections make it easier to hire people of all ethnicities all over the world – creating an international workforce, increasing labour availability and overall skillsets (with different cultures, there are different methods, meaning different skills).

An international workforce means you need **freedom of trade**. This is critical when transporting goods over international boundaries. Freedom of trade is crucial to the existence of companies trading and manufacturing goods all over the world.

Technology has made it cheaper to trade goods and services than ever before. With cheap shipping and international calling – needed for global support teams working in the Tertiary sector – more companies have a global reach than before.

So, technology is the key facilitating factor that provides numerous mechanisms for globalisation to better occur.

As a quick summary:
- **Improvements in Transportation** – larger ships reduce the cost of transporting goods between countries. Also, goods can be transported more rapidly around the world
- **Freedom of Trade** – free trade is becoming more common. This helps to remove socioeconomic barriers between countries, meaning that trading is easier
- **Communication Improvements** – the internet and mobile technology advancements means it is easier for people in different countries to communicate and organise global business
- **Labour Availability + Skills** – other countries (e.g. India) have low labour costs, but high skill levels. There are also less legal restrictions in some LEDCs, meaning that manufacturing industries can utilise the cheap labour costs to drive high profits. By outsourcing their manufacturing, they become globalised

Transnational Corporations (TNCs) are corporations that operate, manufacture and trade all over the world. They have factories and offices in many different countries.

3.1 DEVELOPMENT

Their growth moves more production into LEDCs to reduce labour and resource costs, boosting profits.

Their growth encourages the development of **Newly Industrialised Countries (NICs)**, which have growing consumer markets. This is because people employed by TNCs become able to afford goods and services that companies provide.

Impacts of Globalisation

Impacts at a local, national and global scale

Scale	Positive	Negative
Local	<ul><li>Investments by TNCs create jobs</li><li>These jobs often have higher wages than local businesses</li><li>This gives people a higher standard of living</li><li>Positive multiplier effect on local businesses</li><li>Local people learn new skills</li></ul>	<ul><li>TNCs often make people work long hours in poor conditions</li><li>Profits are sent to MEDCs not local community</li><li>Lack of laws over pollution in LEDCs mean TNCs can pollute the local environment without consequence</li></ul>
National	<ul><li>Most of the products manufactured by TNCs are exported, benefitting the country's economy</li><li>TNCs pay tax to the host country's government</li></ul>	<ul><li>Host countries may become dependent on TNCs, meaning that the government may start to be influenced by TNCs</li><li>People in MEDCs become unemployed as TNCs globalise manufacturing</li></ul>
Global	<ul><li>Global sharing of ideas and cultures. This means people can experience foods and products not otherwise available in their home country</li><li>Increases awareness of events (like natural disasters)</li><li>Highlights global inequality</li><li>Increases awareness of the need for sustainability and of global issues</li></ul>	<ul><li>Operates mainly in the interest of MEDC countries</li><li>Threats to cultures (like to traditions and languages) in countries where TNCs are located. This is because they take their origin country's cultures with them, influencing the local population</li></ul>

THEME 3
ECONOMIC DEVELOPMENT

Case Study – Nike
A transnational corporation and its global links

Nike, established in 1972, is famous for sport equipment and clothing – with their trademark being the globally recognised 'swoosh' symbol. By having factories all over the world, Nike is a good example of a globalised Trans-National Corporation.

Nike's current headquarters are in Oregon, USA. Around the world, they have over 1 million workers working in over 500 factories.

In Nike's early days, products were sold in North America and Europe. Today, China is one of the fastest growing markets for their products.

Many of Nike's factories are now in South East Asia for a few reasons:

- Cheap Labour
- Cheap, local raw materials
- Proximity between SE Asian countries, reducing transport costs
- Free trade
- Expanding markets located in Asia (like in China and India)

China is currently the 2nd largest market for Nike after their 1st largest market, the USA. China also has many Nike factories – partly because of its cheap and abundant labour.

However, because of the 1 child policy, there are growing labour shortages in China. This means that some of Nike's Chinese factories are experiencing hiring difficulties. So,

Image 48: Nike's Headquarters, OR, USA

Nike is moving some factories to South East Asian countries where there is more available cheap labour.

There have been some accusations of:

- Low wages
- Expectations of long hours
- Not enough training
- Poor and unsafe conditions

… that led to some strikes and some factory relocation, meaning people lost jobs.

To improve conditions in some of the places where their factories are located, Nike are trying to control and review:

- Their water and energy usage
- Their pollution
- Their labour policies
- Their methods of waste disposal
- Their social impacts

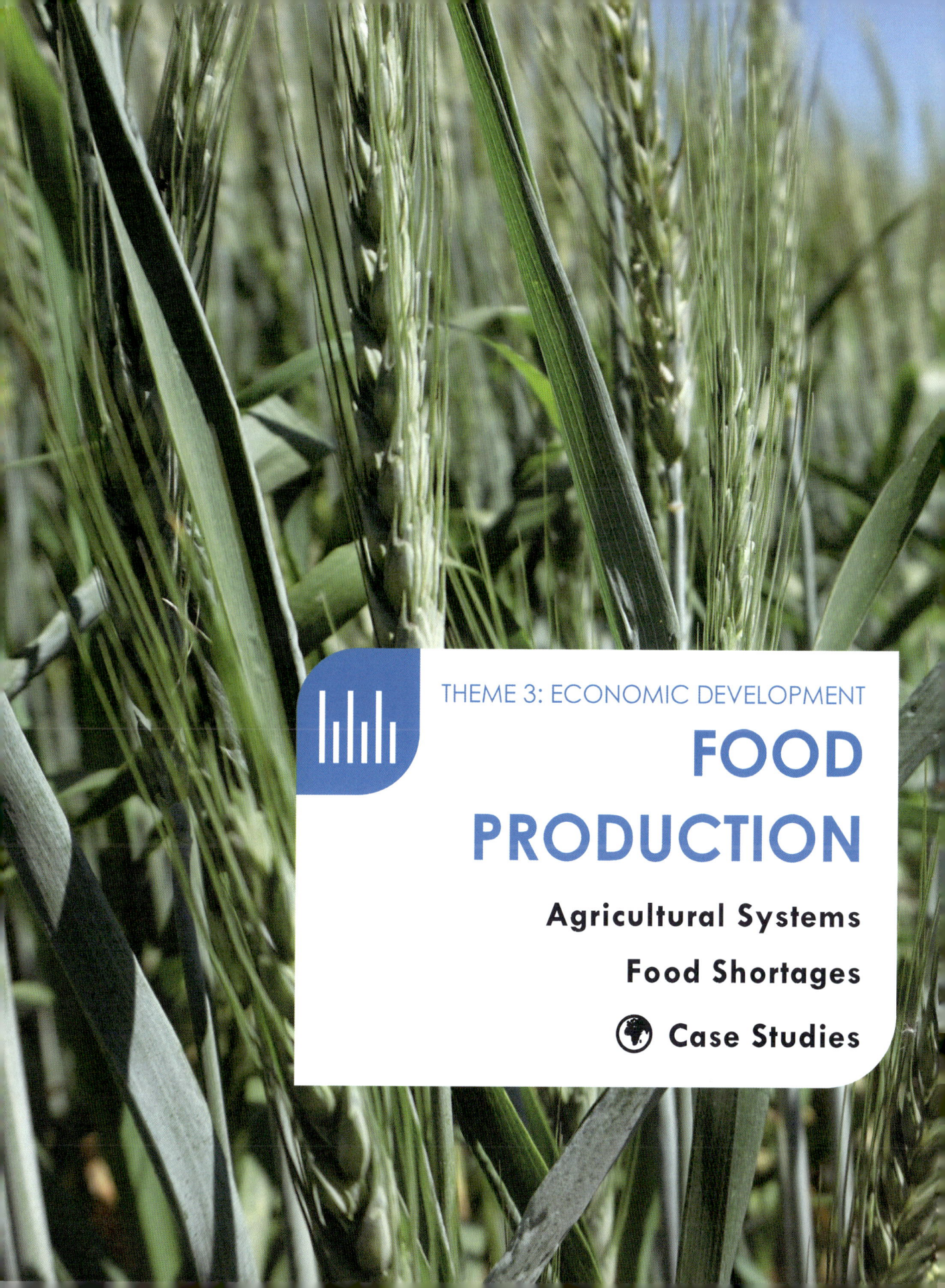

THEME 3: ECONOMIC DEVELOPMENT

FOOD PRODUCTION

Agricultural Systems

Food Shortages

Case Studies

THEME 3
ECONOMIC DEVELOPMENT

> **Describe and explain the main features of an agricultural system: inputs, processes and outputs**
>
> **Recognise the causes and effects of food shortages and describe possible solutions to this problem**

Agriculture is the practice of farming

Types of Farming

> **Farming types: commercial and subsistence; arable, pastoral and mixed; intensive and extensive**

Commercial Farming produces food for sale
Subsistence Farming produces food for the farmer and their family

Intensive Farms use large amounts of money, machines, technology and workers
Extensive Farms use less or smaller inputs and less land

Arable Farms grow crops
Pastoral Farms rear animals
Mixed Farms grow crops and rear animals

Yield is the amount of crop harvested from an area of land. A high yield is important for farmers, as it creates more profit.

Agricultural Systems

> **Describe and explain the main features of an agricultural system: inputs, processes and outputs**
>
> **The influence of natural and human inputs on agricultural land use**
>
> **Including natural inputs (relief, climate and soil) and human inputs (economic and social)**
>
> **Their combined influences on the scale of production, methods of organisation and the products of agricultural systems**

There are 4 main parts to an agricultural system:

- **Inputs** – Things that go into a system
- **Outputs** – Things that come out of a system
- **Processes** – Things that happen to inputs to turn them into outputs
- **Feedback** – The profit being reinvested as an input

3.2 FOOD PRODUCTION

Inputs	Processes	Outputs	Feedback
Physical	**Growing Crops**	Crops	**Profit** to be used as a capital input next year
Climate	Ploughing	Animals	
Relief	Planting	Animal Products	
Landscape	Spraying	Money (Capital)	
Soils	Harvesting		
Sun	**Rearing**		
Human / Economic	Grazing + Feeding		
Workforce	Haymaking		
Farm Buildings	Milking		
Machinery	Shearing		
Chemicals	Calving		
Transport			
Seeds			
Animals + Feed			
Money (Capital)			
Government			
Energy			
Demand			

Agricultural Systems Diagram

The **relief** is the shape of the land. It is easier to plough and work flat land, meaning that arable farms tend to be in flat areas. Pastoral farms are more likely to be in hilly areas than arable farms because animals can roam on land with an uneven relief.

It is also important for there to be good **drainage** to avoid flooding. Most crops and animals can't survive on flooded areas (apart from crops like rice).

The **climate** affects what types of crops can be grown on a farm. Sunny climates provide more energy to help plant growth.

If the **soil** isn't rich in nutrients, the farmer might choose to use fertilisers help to improve crop yield. If it was rich, the farmer could just care for the land, and save money by not using fertilisers to increase yield. Also, if the **soil is fertile** it is more likely for it to be used for arable farming. However, if it isn't fertile it might only be able to support grazing plants (like grass), meaning it is more likely to be used for pastoral farming.

If there is a large demand and there are lots of resources, energy and capital – the farm would be large scale to provide for the large demand.

THEME 3
ECONOMIC DEVELOPMENT

Image 49: A Corn Farm, Iowa, USA

Key Features of Commercial Farming

Commercial Farming is carried out to make a profit.

In MEDCs:
- There are individual, extensive farms where farming is done on a large scale
- There are large inputs (more money is invested)
- There are seasonal workers to reduce the capital input

In LEDCs:
- Less money is invested in the farm
- The farms are run by giant Agribusinesses (huge companies that own land, usually TNCs)

Key Features of Subsistence Farming

Subsistence Farming is carried out to only feed the farmer and their family.

Surplus is sold to the local community at markets. This type of farming is common in LEDCs, where people might not have the money or resources to scale production for commercial farming.

- Moving cultivation
    - There is a small, central village where the farmers are based
    - They farm crops in gardens
    - When the soil quality becomes poor, they move to another area
    - The crops used often grow easily, so less inputs are needed

3.2 FOOD PRODUCTION

Subsistence Farming continued...

- Rice Cultivation is a common form of Subsistence farming, because it has a high yield.
- Little surplus as rice has to support a large amount of the Asian population
- Increases susceptibility to Food Shortages

Food Shortages

Recognise the causes and effects of food shortages and describe possible solutions to this problem

Natural problems which cause food shortages (including drought, floods, tropical storms, pests) and economic and political factors (including low capital investment, poor distribution/transport difficulties, wars)

The negative effects of food shortages; the effects of food shortages in encouraging food aid and measures to increase output

Causes
- **Extreme weather events** – lead to crop failure
- **Natural disasters** – these affect food distribution from damage of transport infrastructure
- **Disease** – farmers can't work, so can't grow food or tend the land
- **Pests** – destroy crops
- **Civil War** – disrupts food supplies; soldiers use up the food supplies
- **Soil Exhaustion** – the soil doesn't have enough nutrients to grow food
- **Enhanced Global Warming** – could lead to crop failure

Effects
- **Cattle Die** – as their food source is lost
- **Malnutrition** – a lack of food or nutrition
    - **Increased susceptibility to diseases** – malnutrition weakens the immune system
    - **People become unable to work** – and perhaps farm, meaning that the famine continues. The family's source of income will be removed, meaning they won't be able to afford food again, further increasing the famine effect
- **Rising Food Prices** – because food is so scarce
- **Starvation** – because food isn't available
- **Mass Migration** – as people flee to try to find food

THEME 3
ECONOMIC DEVELOPMENT

Responses (also effects):

- **An Influx of foreign aid** – the world wants to try to help people in famine
- **Government Efforts to increase food production** – to resolve the source of the famine and to improve the quality of life for the country's residents

Solutions

1. **Green Revolution** – for several decades, scientists have been trying to develop new varieties of disease and drought resistance crops which provide higher yields
2. **Genetically Modifying Crops** (GM) – artificially altering the genetic makeup of crops to allow them to survive in the extreme weather conditions (like deserts) found in many LEDCs. You could then potentially grow crops in places that were previously unsuitable. However, genetic modification is highly controversial because we don't yet fully understand the impact GM crops may have on humans and the environment
3. **Better Technology** – New technologies could allow us to increase food production. By having internet access, farmers can check local market prices (so they know when to sell their crops), the weather (to prepare for weather events) and education resources (so they can learn about new farming techniques to improve their yield). This is especially important in LEDCs, where many people don't have access to the internet
4. **Education** – This can teach farmers to improve their yield, skills and technique. Examples of skills farmers can be taught include:
    a. **Building Dams** – so water can be saved for irrigation
    b. **Using Solar Energy** – this can pump water, meaning that irrigation systems could be used to increase yield
    c. **Building Terraces** – reduces soil erosion and nutrient depletion, so a patch of land can be farmed for longer. This saves money as the farmer doesn't have to buy as much new farming land

3.2 FOOD PRODUCTION

Case Study – Rice Farming, Bangladesh

A farm or agricultural system

Bangladesh has one of the highest population densities in the world. About half of the population are farmers, with rice being one of their **staple crops**.

Major Inputs:

- A hot climate with high precipitation
    - Ideal conditions for rice to grow in throughout the year
- Fertile soils
    - From the silt deposited by the river Ganges in times of flood
- Large population
    - Provides the labour input needed for farming

Image 50: Rice Farm, Bangladesh

The major output is rice, to sell for capital and for the family to eat.

The typical rice farm in Bangladesh is small, meaning rice farming is sometimes for subsistence. However, farms get smaller when land is handed down to younger generations – it is divided up so each person has their own share.

As Bangladesh is an LEDC, many people can't afford machinery. Consequently, most of the farming is done by hand. Farmers are now starting to farm fish in the rice paddies to provide an extra source of income.

During recent floods, some of the crops have been destroyed. This has made it harder to sell large amounts of rice for many people in Bangladesh. Despite this, rice production is still growing due to continued efforts by farmers in Bangladesh, because it is their means of living.

The government has tried to reduce the impact on Bangladesh's rice exports by the recent floods. Funding has been set up to create irrigation systems in drier areas of Bangladesh. These areas would be less likely to flood, meaning that rice can be grown more reliably.

THEME 3
ECONOMIC DEVELOPMENT

Case Study – Famine, Horn of Africa
A country or region suffering from food shortages

NB **Famine**
is when there is not enough food for a given population

In the Horn of Africa area, there has been periodic, recurring famine. There are 2 main causes of this:

Figure 17: Horn of Africa Location

- **War** (Al Shabaab and other Islamic Terror Groups)
    - This prevents food from being transported, meaning that people can't access reliable food supplies
    - The soldiers consume all the available food, leaving nothing for the people
- **Drought**
    - Leads to crop failure, death of cattle and famine. This also increases the food prices, so people can't afford to access food
    - Might provoke further war over resources

The drought led to crop failure in Ethiopia and Somalia (one of the worst hit areas).

Due to conflict in Somalia, people were unable to access reliable food supplies. Because Somalians rely on cattle for food and the cattle died due to a lack of feed, the Somalians also lost one of their main sources of food.

As Somalians don't have access to food, they are more likely to die from Pneumonia and other diseases. When ill, they can't farm the land, meaning that the crops won't recover. If they have a job, illness will prevent them from working and earning money. With no money, they can't afford to purchase food, so become even more ill and so on.

People from Somalia fled to overcrowded refugee camps in Kenya and Ethiopia to access food and to escape from conflict. May die on the journey from starvation and fatigue. When they get there, there isn't enough food to go around (because the camps are overcrowded), meaning more people die.

Organisations like **Care International** are providing aid to famine stricken areas – like Somalia – and refugee camps where the citizens flee to. However, Donor Countries are beginning to contribute less due to concerns over corruption. They worry that the aid they provide won't reach the people in need.

Although some aid is being given, not much has been done to prevent recurrence of the famine. Enhanced global warming is a key source of the recurring drought, and hence, famine.

In Somalia, outside aid has been forbidden by Al Shabaab. As there is no stable government with ongoing civil war, the country is unlikely to recover on its own. This makes Somalia one of the most affected places, as less can be done to help the Somalians.

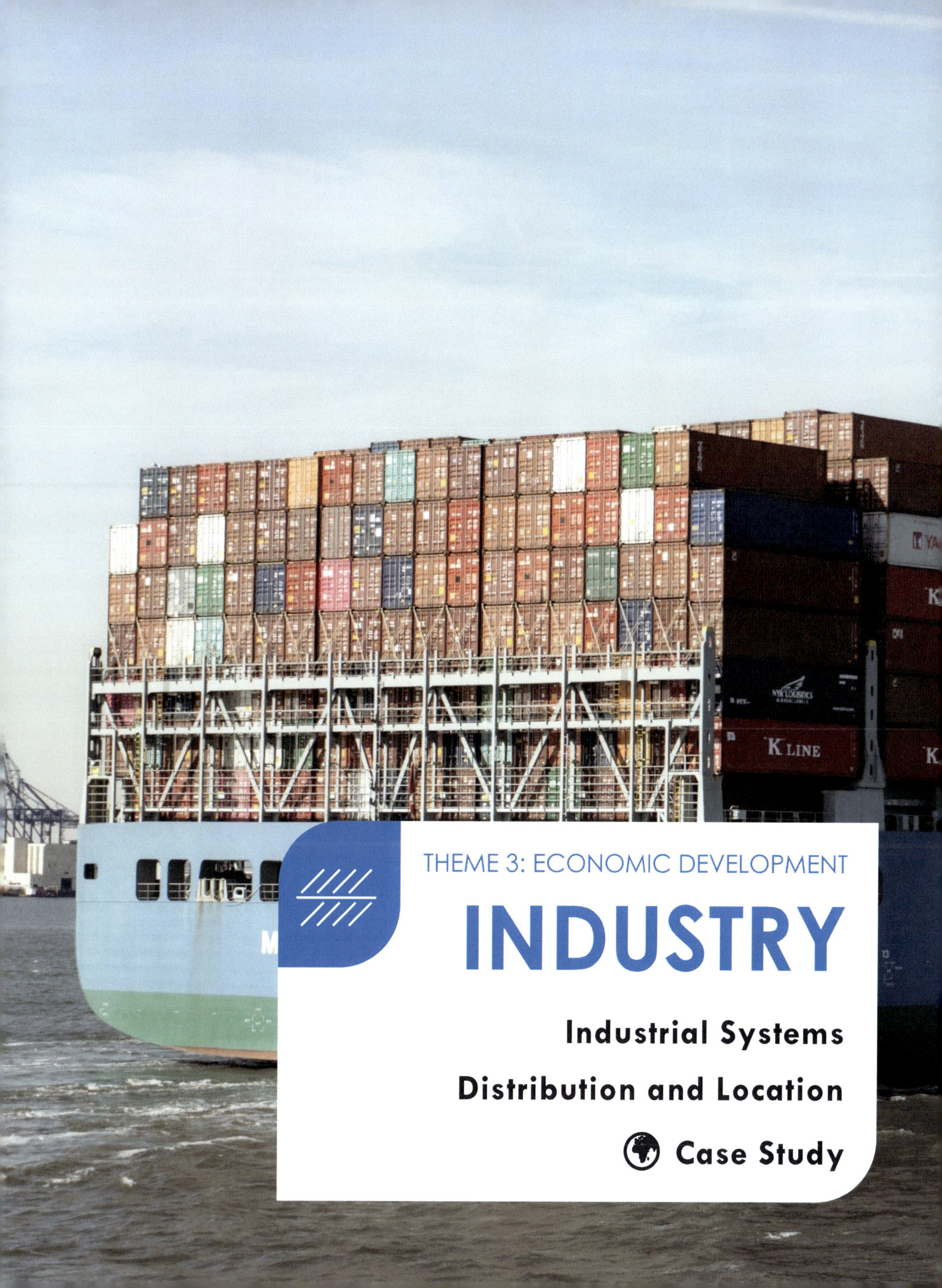

THEME 3: ECONOMIC DEVELOPMENT

INDUSTRY

Industrial Systems

Distribution and Location

 Case Study

THEME 3
ECONOMIC DEVELOPMENT

Demonstrate an understanding of an industrial system: inputs, processes and outputs (products and waste)

Describe and explain the factors influencing the distribution and location of factories and industrial zones

The **Site** is the land that the factory is built on

Industrial Systems

Demonstrate an understanding of an industrial system: inputs, processes and outputs (products and waste)

There are 5 main parts to an industrial system:

- **Inputs** – Things that go into a system
- **Outputs** – Things that come out of a system
- **Processes** – Things that happen to inputs to turn them into outputs
- **Feedback** – The profit being reinvested as an input
- **Waste** – Unwanted things that come out of a system

Inputs

- **Physical**
- Raw Materials
- Natural Features
- Site
- Available Land
- Energy Supplies
- **Human / Economic**
- Capital (Money)
- Technology
- Labour
- Transport
- Markets
- Policies
- R&D

Processes

- **Processing of Raw Materials**
- Printing
- Packaging
- Assembling
- Programming
- Sewing
- Painting
- Assembling

Outputs

- Products
- Capital (Money)
- Waste

Feedback & Waste

- **Feedback**
Profit to be used as a capital input next year
- **Waste**
Recycled to provide more inputs
Burnt to provide energy

Industrial Systems Diagram for Manufacturing Industries

3.3 INDUSTRY

Distribution and Location

Describe and explain the factors influencing the distribution and location of factories and industrial zones

The influence of factors including land, labour, raw materials and fuel and power, transport, markets and political factors

Their combined influences on the location, scale of production, methods of organisation and the products of the system

Industrial zones and/or factories with respect to locational and siting factors

- **Raw Materials** – As these are often quite big and heavy, they cost lots to transport. This means that many industries locate themselves near the raw material or near good transport links (like rail networks) to reduce transportation costs
- **Power** – Industry needs lots of power. In LEDCs, power often isn't widely available, so many industries in LEDCs are located near power stations or primary distribution grids close to those power stations

Image 51: Transport is an Important Factor when locating industry - FCCA Freight Train, Peru

- **Fuel** – Some industries rely on their own sources of power to heat furnaces. They sometimes do this by burning fuel. This means that these industries are more likely to be located where this fuel is available or close to good transport links (like rail, to import fuel)
- **Natural Landscape** – Features, like mountains, make it difficult to build the infrastructure needed to transport goods. Industry is consequently more likely to be located on even terrain than on mountainous terrain
- **Land for Site** – As Factories tend to be large in size, to be cost effective they often locate where cheap, flat land is available
- **Government Policies** – These might encourage industry to locate in a certain area by providing capital incentives. This means that industry may be more concentrated in areas where such policies apply
- **Market** – Industries make goods to sell in markets. So, it makes sense for industry to be in the market (or near it) to reduce the cost of transporting finished goods
- **R&D** – Research and Development is more likely to occur in MEDCs. This means that more high-tech industries start in MEDCs as they require certain education levels and technologies to develop new products

153

THEME 3
ECONOMIC DEVELOPMENT

- **Labour** – Factories need to hire people to perform tasks to operate the factory as well as to service and run the machines. This means that it is more likely for an industry to locate in an area with cheap, appropriately skilled an abundant labour in order to minimise the cost of labour in the finished goods or services
- **Transport Accessibility** – Industries need to move materials and goods to and from factories. Also, workers need to commute to the factory. This means that many factories are located near major, uncongested roads so heavy goods can be transported and so workers can more easily commute to the factory. Additionally, factories may be located near ports or train stations to make it easier to transport goods globally
- **Local Ambience / Environment** – Pleasant working conditions or good local facilities (like leisure facilities, cafes and shops) make workers want to work at the factory or live near it. This is why some factories are located near pleasant areas, so workers can live more comfortably before commuting to their workplace.

Types of Industry

Industry types: manufacturing, processing, assembly and high technology industry

Primary Sector	Secondary Sector	Tertiary Sector	Quarternary Sector
• Collecting Raw Materials • Farming • Mining	• Manufacturing • Automotive • Technology - e.g. Computers / IT equipment	• Providing a service • Banking • Teaching • Lawyers	• Knowledge Based Jobs • Research • High Tech Industry (e.g. Google)

Manufacturing, Processing and Assembly Industries are Secondary Sector Industries. This is because they process raw materials to create new products. They are more common in LEDCs than MEDCs, which could be because they require a lower level of skill or education – caused by a lack of freely available education. LEDCs also may not have the technology or access to communication to provide Tertiary Sector services.

Additionally, as the cost of transport, power and raw materials is now lower globally, labour costs become a determining factor in an industry's location. This is perhaps why these 3 secondary sector industries are more frequently located in LEDCs, because of the cheap and abundant supply of labour there.

Some factors that influence the location of Secondary Industries:

- **Transportation and Accessibility** – goods and raw materials are heavy
- **Land** – factories for these industries take up lots of land
- **Labour** – needs to be cheap to reduce costs

3.3 INDUSTRY

High Technology Industries involve the application of technology into a product. This could be a computer or mobile phone. As these devices are smaller, they require less raw materials to be manufactured. This means that the cost of raw materials and transport is lower, so they are no longer limiting factors to the industry's location. High tech products sell for high prices, give a high capital output and don't cost much to transport (because they are normally very light). This means that other location factors determine the industry's location.

As these high technology industries use advanced technology, lots of Research and Development (R&D) needs to take place to develop new technologies. As R&D requires a high skill and education level, these industries employ university graduates. These graduates often desire to work in a pleasant environment, meaning that many high tech industries locate in attractive areas with good facilities. Evidence of this is the high concentration of high tech industries located in Silicon Valley.

However, lots of high tech industries are also located in Asia (like Samsung), often in science and research parks. These contain facilities which attract university graduates to work in the high technology industries.

Image 52: Silicon Valley, USA

(See Next Page for Factors influencing the location of high tech industries)

155

THEME 3
ECONOMIC DEVELOPMENT

Some factors which influence the location of high tech industries:

- **Education Level** – high tech industries use advanced concepts and processes in manufacture, testing and design
- **Transport** – needed to export goods and for employees to commute
- **Abundant, Cheap Labour** – for maintenance and assembly on high tech production lines
- **Good Local Environment** – attracts graduates
- **Science Parks** – so high tech industries can collaborate
- **Cheap, Spacious Land** – to build factories, offices and to give the industry potential for growth

NB The type of industry affects what processes occur in the industrial system.

Case Study – Toyota, Burnaston, UK
An industrial zone or factory

NB Burnaston is close to Derby

Toyota is a car manufacturer, and opened their Burnaston factory in 1992.

But why did they locate it there?

- **Transport** – Located on a major road junction, with easy transport links to the rest of the UK (e.g. M1). The plant is also near the East Midlands International Airport, which facilitates importing lightweight car electronics by plane. It is a major hub in the UK for international cargo. Thus, these 2 links are crucial for transporting raw materials and cars
- **Market** – The UK has a large population (of over 65 million), meaning there is a large domestic market for Toyota to sell cars to
- **Energy** – The UK's national grid power system means that Burnaston has a reliable supply of electricity to power manufacturing machinery
- **Communications** – Excellent nationwide broadband and mobile coverage means that it is easy for the factory remain in close contact with Toyota's Japanese headquarters
- **Government Policies** – The British Government wanted Toyota to invest in

Image 53: Toyota Factory, Burnaston

3.3 INDUSTRY

the UK and helped Toyota to find employees. The government also maintains the adjoining A50 and A38 roads, meaning that transport can be kept fluid

- **Education** – The nearby universities (like Derby University) can supply well educated graduates who could work in Research and Development at Toyota Burnaston
- **Labour** – Derby is a traditional location for manufacturing industries (Rolls Royce being a classic example), meaning that there are already lots of high skill workers in the area
- **Local Environment** – There are lots of recreational facilities (like Cinemas and Shopping Centres) nearby, along with good hospitals (like the Royal Derby) to provide all the services that workers could need
- **Relief / Natural Landscape** – The site the factory is built on is flat, meaning construction is easier. There is lots of free land for future expansion

Inputs	Processes	Outputs
• £1.15 billion investment • Over 2,500 workers • Many suppliers for the factory are in the UK or easily accessible by close transport links • The factory site is approximately 600 acres • Investment from UK government • Integration – English is Japan's second language • Access to European market • Near Toyota's Chester engine factory • Derby is traditionally a manufacturing city	e.g. • Welding • Painting • Assembly • Pressing	• Toyota Avensis and Auris cars • Most are sold within Europe, reducing transport costs • Profit • Waste

THEME 3
ECONOMIC DEVELOPMENT

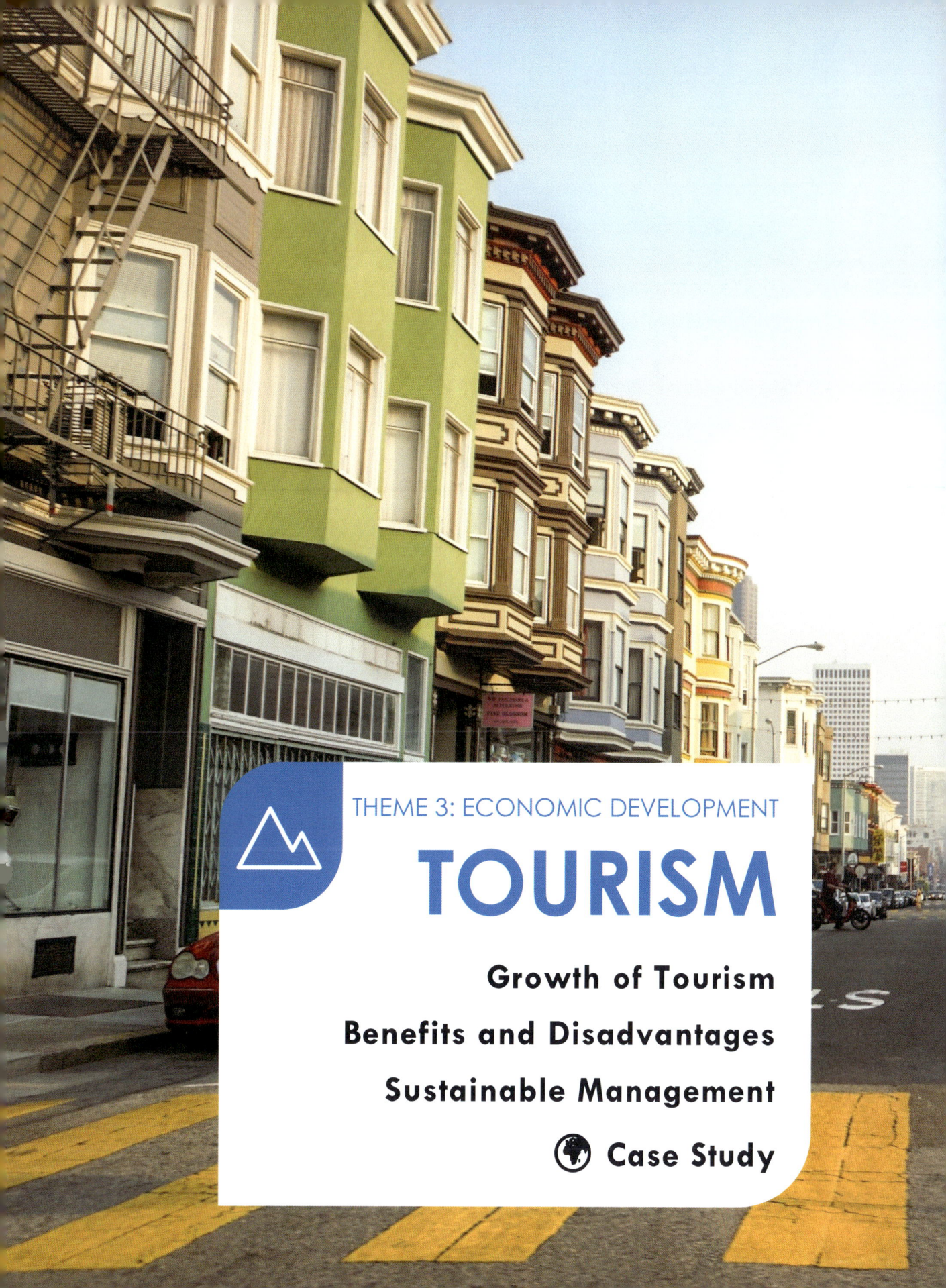

THEME 3
ECONOMIC DEVELOPMENT

> **Describe and explain the growth of tourism in relation to the main attractions of the physical and human landscape**
>
> **Evaluate the benefits and disadvantages of tourism to receiving areas**
>
> **Demonstrate an understanding that careful management of tourism is required in order for it to be sustainable**

Tourism is the industry that provides services to tourists (like accommodation, flights, tours and reservations)

An **International Tourist** goes on holiday to another country, whilst a **Domestic Tourist** goes on holiday within their home country

A **Package Holiday** is a bundle of flights and other services

A **Budget Airline** is an airline that has cheap flights. They make flights cheaper by removing all luxuries, like hold luggage and food / drinks, from the base fare and by charging extra for their inclusion

Growth of Tourism

> **Describe and explain the growth of tourism in relation to the main attractions of the physical and human landscape**

Bern, Switzerland

Reasons for Tourism Growth:

- **Leisure Time:** Most workers have a 2-day weekend with several weeks' worth of holiday leave. This time can be spent on leisure activities, like going on holiday.

- **Paid Holiday:** An increasing number of workers still get paid whilst on holidays, encouraging them to take their holiday time, as they will not be economically hit when doing so. By encouraging workers this way to take holiday, the company attains a healthier, more rested and more motivated workforce.

- **Income:** People working in higher paid sectors and jobs have a higher disposal income – the number of which is on the rise. This means that more people can afford to go on holidays.

- **Transport:** Travel has become cheaper, and there are now more domestic airports, which enables travel with less transit time. New airplane routes also allow people to travel to new destinations further afield.

3.4 TOURISM

- **Advertising:** As a society, we are increasingly exposed to the promotion of holidays and destinations through a variety of different media outlets. This increases awareness of holiday destinations and travel promotions, increasing the likelihood of people holidaying to these destinations.

- Travel Media: There is a large selection of travel TV, blogs and magazines for people to read. This means that people are encouraged to consider travelling to places they hadn't previously heard of.

- **Tourist facilities:** Facilities for tourists have improved and increased in recent years. There is now a greater selection of hotels and restaurants, permitting for better fulfilment of people's needs. Also, many facilities have become standard, like Wi-Fi and electricity in tourist accommodation.

- **Freedom:** More people are free to travel without visas or travel permits due to changes in government policies. This increases the number of potential tourists.

Krakow, Poland

- **Range of holidays:** There is now a wider selection of holiday types to cater for the specific needs of different tourists. This increases the variety of tourists going on holiday, and tourism's market as a whole.

- **Ease of Booking:** The internet has simplified the process of booking a holiday. Package holidays eases the booking process even further, with the tourist only having to make 1 booking in 1 place. This boosts the tourism sector because it makes preparing to go on holiday much simpler, meaning people are more likely to book bigger, more expensive holidays.

- **Credit Cards:** Credit cards make it easier to book holidays and pay for goods abroad.

- **Passport Ownership and Visa Regulations:** More people own passports, and internet makes it simpler to apply for visas. This means more people are free to travel to places they couldn't go to before.

- **Retirement and Life Expectancy:** More people are living healthily for longer. Also, an increasing number of people retire with a pension. This means that more retired people have the funds and health to go on holiday.

Reasons for Tourism Growth in LEDCs

- **New Destinations:** People are getting bored of traditional destinations and want to try new, untraveled places.
- **Exchange Rates:** LEDCs tend to have weaker currencies, making it cheaper to go on holiday there.
- **Advertising/Ease of Booking:** Because the internet is so accessible, LEDCs can advertise themselves across the globe. The internet also makes it easier than ever before to book holidays to them.
- **Transport:** Many LEDCs have recently upgraded their airport capacity, meaning they can handle greater amounts of tourism
- **Security:** More LEDC countries have stable governments with less security issues. This means people feel more comfortable travelling there.

Tourism decline may occur because of:

- **Terrorism** e.g. Terror Attacks and Terrorist threats
- **Crime** e.g. Mexico
- **Natural disasters** e.g. tsunamis, hurricanes, tropical storms
- **Economic downturn** e.g. recessions and debt crisis in Greece
- **War** e.g. some parts of the Middle East and Africa

Attractions of The Physical and Human Landscape

- **Coasts** have sun, water sports and beaches. These allow people to have winter sun and experience a wide range of enjoyable leisure activities. The growth of cheap flights from Europe to destinations like the Canary Islands in Spain means that holidays to the beach are more accessible than ever – growing the coastal tourism industry
- **Mountains** provide a varied, interesting landscape as a stunning backdrop to a holiday. Snow in Winter, and high-altitude sun in Summer, means that the mountains are host to a variety of sports and leisure activities (e.g. skiing, hiking and mountain biking) throughout the year
- **Cities** have history, culture and shopping facilities that people could spend their money on. Additionally, cities like San Francisco and London have unique cultures, history and experiences which can't be found anywhere else. Hence, people may be willing to travel long distances to have these experiences.

3.4 TOURISM

Benefits and Disadvantages
Evaluate the benefits and disadvantages of tourism to receiving areas

Benefits	Disadvantages

Social
 • Local People can share their traditional culture and heritage. By keeping old cultures active, the cultural richness of an area is maintained
 • Tourists could visit local galleries which showcase local culture and talent
 • Tourists could pay to visit local monuments, meaning that the government can afford to maintain them
 Economic
 • 'Tourist Tax' can give the government money to invest in local services (e.g. the Balearic Islands' Ecotourism Tax)
 • Enhanced employment opportunities because tourism creates jobs in the service industry (like in Hotels and Cafes)
 • Infrastructure may be upgraded to give the tourist a better experience in the country – additionally benefiting the local people and country
 • People in the tourism industry may learn new skills which can be repurposed in local industries
 Environmental
 • Increased awareness over conservation, if tourists respect and enjoy the local environment
    o In turn, this may lead the creation of national parks
    o This increases the value of the natural environment, meaning that tourists may want to help protect it
 • The money gained from tourism could be directed into conservation projects by the government
	Social
	 • Increased tourism could lead to greater congestion on a country's road network, damaging the local people's quality of life
	 • Tourism may increase crime rates
	 • Racial tensions could arise between prejudiced tourists and local minorities
	 • Large resorts might be built on sacred land, destroying cultural heritage
	 Economic
	 • Farmers (or other food providers) might lose their jobs, as hotels or chain restaurants may serve imported food to tourists
	 • Locals may get the low paid jobs, whilst foreign workers may receive higher paid jobs – which is unfair for the local workers
	 • Some jobs are seasonal, so people might not have a consistent income during the year (e.g. Ski resorts). So, people may have multiple employers
	 • Tourism is an unstable industry that is affected by external factors, so people might lose their jobs if the destination become undesirable to travel to. This could happen after a natural disaster or terrorist attack
	 • Most of the profit from tourism is taken by TNCs, as their chains often control the tourism industry (like hotel groups)
	 • Many low skilled jobs in tourism have low salaries
	 • Pressure could be put on the country's infrastructure
	 Environmental
	 • Destruction of natural habitats to construct hotels, airports and roads
	 • Pollution from tourists' litter, nightlife and light pollution from floodlit hotels
	 • Issues like boats damaging coral reefs or erosion of cliffs from heavy footpath use

THEME 3
ECONOMIC DEVELOPMENT

Sustainable Management
Demonstrate an understanding that careful management of tourism is required in order for it to be sustainable

Sustainable Tourism – activities or services that are sustainable in a social, economic and environmental sense

Image 54

Tourism can damage the environment. A couple of examples include CO_2 emissions from transport and the destruction of natural habitats in the construction of tourist resorts. Tourist resorts have other impacts too: the energy sources of many resorts involve non-renewables; the transport of food and goods produces CO_2; and waste is sometimes dumped.

Ecotourism aims to lower the impact of tourism on the environment by:
- Minimising the impact of transport
- Running conservation programmes
- Helping local people financially and socially
- Raising awareness to tourists about the local environment
- Encouraging respect to the environment and the locals' culture

Ways this can be done:
- Using renewable sources of energy
- Employing locals in hotels and restaurants instead of foreigners
- Recycling waste
- Encouraging tourists to engage in the local culture
- Using local products where possible (this could be food or fuel)
- Educating locals and tourists about the importance of conserving the local environment

However, there is some criticism:
- Tourists still have to travel to tourist destinations, giving out CO_2 emissions
- Guests may not follow the rules established to keep the tourism sustainable and environmentally friendly

3.4 TOURISM

Case Study – Iceland
An area where tourism is important

The Land of Fire and Ice

Iceland, the Land of Fire and Ice, has many beautiful and incredible features that could attract tourists. Barren and wild volcanic and glacial landscapes give it an otherworldly feel. From the active Volcanoes, to the immense Glaciers, Iceland is truly a tourist magnet.

Since the last decade, the increased popularity of stopovers on transatlantic flights in Iceland means that more tourists are visiting the country than ever before. Consequently, tourism in Iceland is becoming a key part of the country's economy, with over 1 million tourists arriving in Keflavik each year! The number of tourists in Iceland each year is over triple Iceland's current population.

The huge growth in the number of tourists, and the market for tourism, has created jobs in shops, hotels and other tourist facilities. As 'Volcanism' doesn't have peak or off seasons, there is a steady stream of tourists visiting Iceland for its volcanoes throughout the year – giving the locals a reliable source of income.

In 2016, 33.9% of Iceland's export revenue was from tourism, with 62,500 tourism-related jobs. Also, in 2016, nearly 1.8m people visited Iceland, nearly treble that of 2010.

The glaciers have prompted the growth of companies offering glacier tours, kit shops, hotels, information centres and cafes.

The volcanoes have provoked the growth of volcano tour companies, lava tube cave tours and the popularity of bathing in the Blue Lagoon.

Iceland's rare blend of volcanic activity and glaciology is due to its situation on the Mid Atlantic Ridge – a constructive plate boundary. This characteristic alone gives people the opportunity to go to Thingvellir for a truly awesome swim between 2 tectonic plates.

Attractions Overleaf

165

THEME 3
ECONOMIC DEVELOPMENT

Attractions

- **Vatnajökull glacier** – the largest in Europe. Tourists flock here to get a taste of what the ice age might have been like
- **Golden Circle Route, South Iceland** – this connects Thingvellir, Gullfoss and the Geysir Geothermal Area, all beautiful natural attractions
- **The Blue Lagoon** – a geothermal pool with hot, silica rich waters. It allows tourists to experience the unique juxtaposition between hot waters and a cold, aurora lit night sky
- **Icelandic Churches** – beautiful buildings which give a cultural insight into traditional Icelandic life
- **Harpa** – a cultural centre which showcases modern Icelandic culture
- **Geothermal Energy Exhibition at Hellisheiði Power Plant** – allows visitors to understand the use and generation of Geothermal power in Iceland

The Geysir Geothermal Area, Iceland

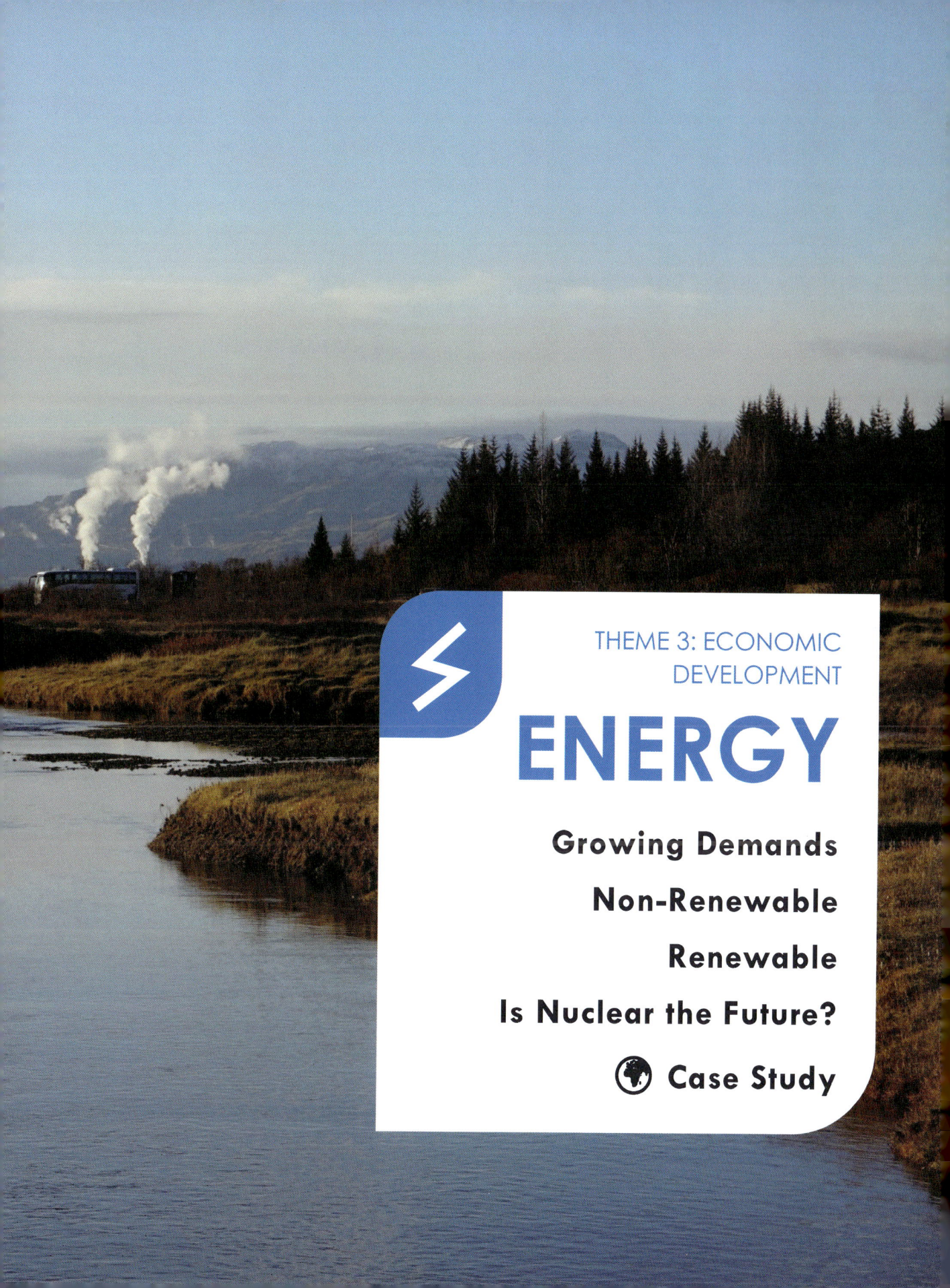

THEME 3: ECONOMIC DEVELOPMENT

ENERGY

Growing Demands

Non-Renewable

Renewable

Is Nuclear the Future?

🌍 Case Study

THEME 3
ECONOMIC DEVELOPMENT

Describe the importance of non-renewable fossil fuels, renewable energy supplies, nuclear power and fuelwood; globally and in different countries at different levels of development

Evaluate the benefits and disadvantages of nuclear power and renewable energy sources

Growing Consumption

World Energy consumption has vastly increased since 1990 and is set to further increase in the future. Today it is over 50% greater than in 1990, increasing by around 5% per year.

But Why?

- Technology requires energy
- Increased industry in LEDCs and **NICs** (like India and China) increases global energy demand
- Economic Development and Increased population
    - An increased population with more economic development leads to more users of technology, which requires more electricity

Both problems and benefits are created by our growing energy consumption:

Problems	Benefits
<ul><li>We are using up non-renewable stores</li><li>The burning of fossil fuels causes enhanced global warming and increased air pollution</li><li>Conflict over Resources</li><li>Safety, like dams bursting</li></ul>	<ul><li>Makes life easier<ul><li>Electrical Appliances</li><li>No need to collect firewood anymore</li></ul></li><li>Transport is now much easier</li><li>Industry can grow (industry needs energy), growing the economy</li></ul>

Over time, our use of renewables, coal and natural gas has steadily grown, scaling to match our increased energy demands. Our usage of nuclear energy hasn't increased that much, mainly because the technology is still being developed and concerns over how to protect the environment long term from nuclear waste.

Our use of **renewable energy resources** is steadily growing, due to a reduction in installation costs. However, at the moment, we can get more energy out of **non-renewables** than renewables; the facilities needed to exploit non-renewable energies also already exist (in MEDCs), meaning that they can provide lots of cheap power. Cheap power is key to the economic development of NICs, so as it is often cheaper to build a non-renewable power station that leverages large reserves of fossil fuels than building a renewable energy facility, many NICs still rely on fossil fuels for generating much of their power.

3.5 ENERGY

Non-Renewable Fossil Fuels

Non-renewable fossil fuels including coal, oil and natural gas.

Non-Renewable Energy – Fossil Fuels				
Type	Where?	Information	Advantages	Disadvantages
Oil	MEDCs	Formed by compressed, decomposed, prehistoric plankton. Extracted from a borehole by oil rigs. It is then refined.	• Easy to transport • Only fuel used (on a large scale) for transportation • Less polluting than coal • It is a raw material in chemical industries	• Greenhouse emissions • Oil Spills cause Pollution • Oil rig work is dangerous • Only a few countries have oil resources, so the market is controlled
Coal	LEDCs	Sedimentary Rock formed from prehistoric trees growing in tropical swamp forests. It is mined underground or accessed by quarrying	• Cheap – makes it ideal for use in LEDCs • Existing technology to use it to generate power	• Air pollution • Visual pollution from mining • Noise from mining • Dangers to mines • Subsidence
Natural Gas	MEDCs; LEDCs	Formed like oil or from swamp plants. Extracted from a borehole. Can be obtained by Fracking – which is very controversial.	• Easy to transport • Less polluting than coal • Cheaper than oil • Cheaper to build gas powered power plants than other types of fossil fuel power plants	• Greenhouse emissions • Gas rig work is dangerous • Only a few countries have natural gas resources, so the market is controlled • May become increasingly expensive when the accessible supplies are depleted

THEME 3
ECONOMIC DEVELOPMENT

Renewable

Renewable energy supplies including geothermal, wind, HEP, wave and tidal power, solar power and biofuels

Renewable Energy				
Type	**Where?**	**Information**	**Advantages**	**Disadvantages**
Wind	MEDCs	A large number of turbines located in an open space produce power. As the wind rotates the turbine, power is generated.	• No air pollution • Little effect on the local ecosystem • Strong winds in winter, meeting peak demand • Cheap energy production (wind is free)	• Can't be used in calm or in storms • Visual pollution • Inefficient – lots of turbines are needed to generate enough power
Solar	Sunny Areas + remote locations in MEDCs	Solar panels convert light into energy. Used by individuals, as the power output fluctuates too much to be connected to the national grid	• Safe • No pollution • Cheap energy production • Effective for low power use (like for heating) • Most useful in sunny places and isolated areas in LEDCs	• High Initial cost • Not as effective in cloudy areas • Less effective for high power use cases (like TVs) • Less useful at high latitudes – because there are shorter days in winter, a period of peak demand
Geothermal	Volcanic Areas; LEDCs; MEDCs	Extracted from hot rocks. Water is pumped through rocks, becoming hot. This hot water turns to steam, powering the turbine. Hot rocks are closer to surface at	• Extremely cheap • No greenhouse gases are released • Water is reused • It is a reliable, continuous source of energy • Power production is unaffected by the weather	• Restricted to areas of tectonic activity • Risk of volcanic eruptions • Groundwater is often saline or poisonous, as minerals are absorbed by the water from rocks • Each well is often only used for 25 years

3.5 ENERGY

		plate boundaries		
Hydro-electric Power Stations (HEP)	LEDCs (their main source of renewable energy)	A dam built across a river, holding back the water – storing it in a reservoir. This water is released through a turbine to generate power	• Energy can be continually produced • Can respond to demand (you can increase the flow through the turbine) • No fuel costs • No air pollution • Leisure potential in the reservoir • The reservoir is a water source	• Expensive to build • Flooding destroys habitats • People are displaced when the valley is flooded to form the reservoir • Collapse of old dams is dangerous • Alters the groundwater level • Traps sediment • Disputes over who owns the river
Wave	MEDCs	Wave Turbines harness the energy from the sea from strongly breaking waves	• No pollution • Low or free running costs • Waves are predictable	• Expensive to maintain and construct • Wildlife Hazard • Dangerous to boats • Visual pollution
Tidal	MEDCs	Tidal Barriers produce energy in areas with strong tides	• Produce loads of power • Tides are predictable	• Loss of river estuary habitats • Visual pollution
Biofuels	MEDCs; LEDCs (on a small scale)	Come from biomass broken down by bacteria	• Prices could be more stable than fossil fuels • Fuel could be sourced locally • Less pollutants released when burnt than fossil fuels • **Carbon Neutral**	• Land where food was previously grown is now used to grow biofuel crops, raising food prices

171

Fuelwood

This is more commonly found in LEDCs, mostly because of how cheap it is, and how people can get it locally.

Advantages:

- It is free for the user
- No advanced technology needed to burn it – making it an accessible energy resource
- It can be sustainable if the fuelwood used is replanted
- It can be sold for a profit

Image 55: People often have to travel long distances to gather fuelwood

Disadvantages:

- In some places, more is being used than grows back. This makes it unsustainable
    - This means that local supplies are depleted
    - So, people must travel further to get their fuel
        - This means that children may not be able to go to school if they have to travel further to collect fuelwood
- Deforestation to collect fuelwood causes some problems
    - Soil Erosion
    - Mineral Depletion of Soil
    - This could lead to desertification
    - This would lead fuelwood becoming scarce and unavailable
- Air Pollution from burning fuelwood as it is often inefficiently burnt

Image 56: Nuclear power plant, Cattenom, France

Nuclear Power – The Future?

Nuclear power uses radioactive uranium and plutonium recovered from spent uranium or decommissioned nuclear weapons to produce energy. This isn't renewable because our supplies of uranium will eventually run out. However, compared to other fuels, very little uranium is needed to produce large quantities of electricity. MEDCs like the USA and France use uranium to produce a large portion of their power.

3.5 ENERGY

However, radioactive waste – produced by nuclear fission – is very dangerous, so needs to be safely disposed of. We do this by burying it deep underground where it can't harm anyone or anything. Despite nuclear fission having its dangers, many countries still use it to produce energy.

But where do we put Nuclear Power Plants?

- Mainly in **MEDCs** because Nuclear Power involves advanced and expensive technology. Countries like Japan often use nuclear power, because they don't have any local supplies of fossil fuels. Countries like the UK use nuclear power to supplement our other methods of power generation, meeting the demands of the consumers
- Near water supplies, like the sea, for cooling the very hot reactors
- On flat, cheap land – nuclear power stations are very large
- Far from settlements – due to safety concerns over nuclear leaks or problems caused by natural disasters near power plants – a few examples including the Fukushima event in 2011 and the Chernobyl event in 1986

Compared to fossil fuels, it is a cleaner alternative to other non-renewable energy sources.

Advantages	Disadvantages
- Very little uranium is needed to produce colossal amounts of energy – much less than the amount of oil needed to produce the same amount of energy - Doesn't contribute to acid rain or enhanced global warming - Research is being done to address our safety concerns - Many governments support the construction of clean energy sources, like nuclear power plants - Safety measures are put in place (like the mandatory use of radiation dosage badges) to reduce any potential risks - Could reduce our use of, and reliance on, fossil fuels	- People have concerns over the safety of nuclear power - Large scale accidents are possible – potentially from natural disasters - People living right next to a nuclear power plant could get radiation sickness - Nuclear power plants are very expensive to build and decommission - Nuclear waste is long-lasting, dangerous and hard to dispose of (never mind the cost of maintaining the disposal sites)

THEME 3
ECONOMIC DEVELOPMENT

Case Study – Iceland
Energy supply in a country or area

Globally, renewable energy only makes up a small percentage of energy production. However, Iceland is in a unique position where over 66% of its energy comes from geothermal power plants, and over 15% from Hydro-electric Power Plants (HEP Plants). This means that over 81% of Iceland's energy comes from its own renewable and sustainable energy resources. The rest of its energy comes imported fossil fuels for cars and ships. Iceland is the only western country where all the energy produced is from sustainable, emission-free sources.

Iceland is uniquely situated on the Mid-Atlantic Ridge. This is a constructive plate boundary between 2 plates (the Eurasian and North American), moving apart by a tiny amount every year. There is also a high level of precipitation in Iceland. These 2 factors mean that Iceland has a large potential for geothermal and HEP energy production.

Most of Iceland's population lives in the south west, where many geothermal energy sources are found (like on the Reykjanes peninsula).

The other geothermal hot zones are in a diagonal band across Iceland.

HEP power plants are built on glacial rivers, fed by melt water, and geothermal plants near clusters of high temperature rocks.

Image 57: HGPS, Iceland

Hellisheiði Power Station (HGPS) is the 6th largest, in terms of electricity production, in the world – producing around 300 MW of electricity. It was built in 2006, and is in the south-west of Iceland. It provides electricity and hot water for the Reykjanes area, including the capital, Reykjavik. HGPS is run by ON Power. However, the power plant has brought some sulphur dioxide pollution, from beneath the ground, to the surface.

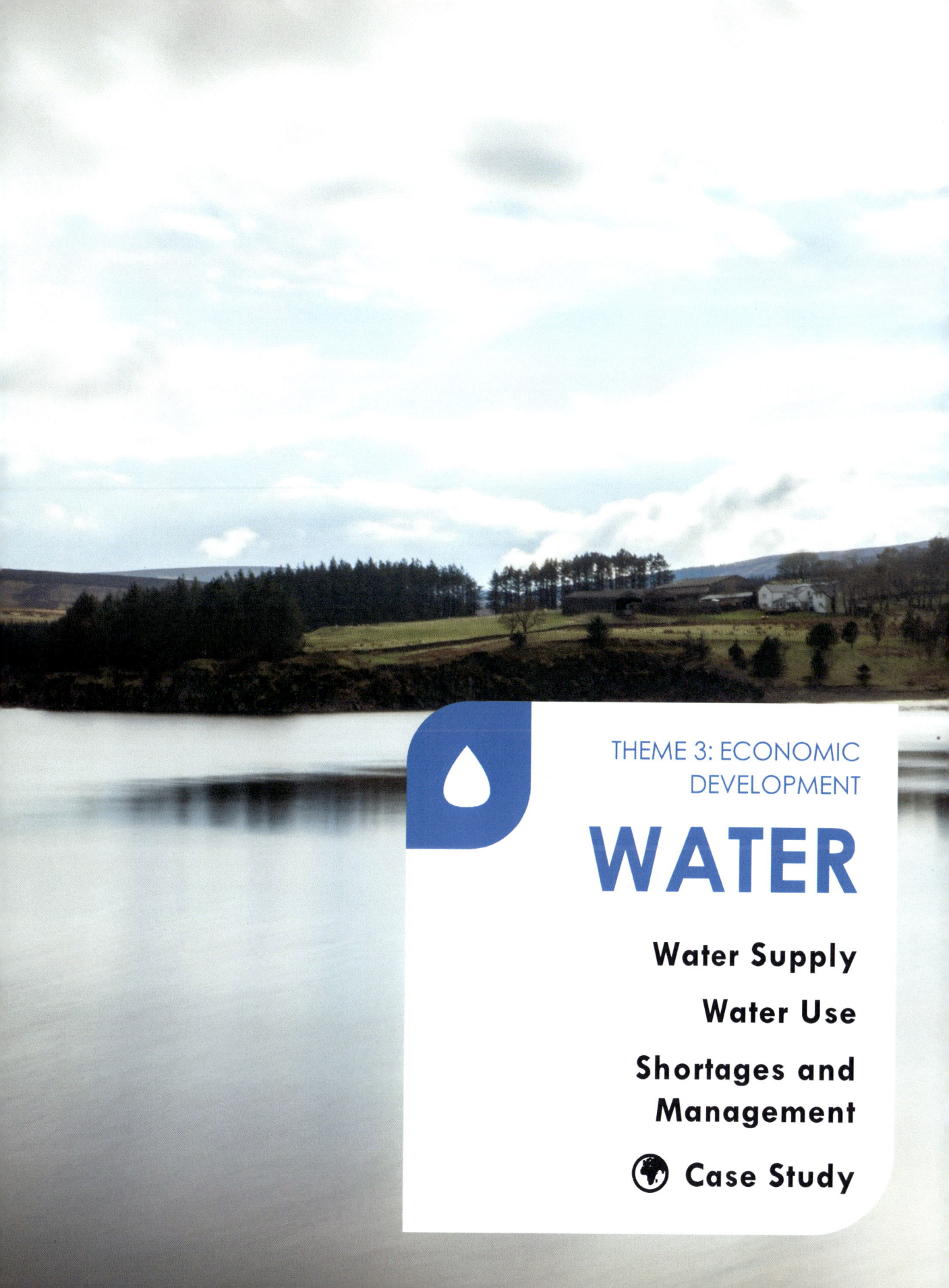

WATER

THEME 3: ECONOMIC DEVELOPMENT

Water Supply

Water Use

Shortages and Management

Case Study

THEME 3
ECONOMIC DEVELOPMENT

Describe methods of water supply and the proportions of water used for agriculture, domestic and industrial purposes in countries at different levels of economic development

Explain why there are water shortages in some areas and demonstrate that careful management is required to ensure future supplies

Water Supply

Methods of water supply (including reservoirs / dams, wells and bore holes, desalination)

Surface Water is water from rivers and lakes. River flow is variable, so rivers are usually dammed to collect water in **reservoirs**. This water is then piped to consumers. Often, reservoir sources are used in dry periods when rainwater isn't available.

Groundwater is water from the porous holes in permeable rocks (like Sedimentary Rocks). Water-containing layers of rock are called **aquifers**. Water can be extracted from these porous holes by digging wells or drilling boreholes into the rock. The water is then brought up to the surface by a pump. We need to make sure that we don't exploit aquifers, so we shouldn't take more than can be naturally replaced. Aquifers don't lose water content through evaporation, as they are underground.

Desalination is when sea water is distilled to remove the salt content, so we can use and drink it. However, this is very expensive (because large quantities of energy is required for the extraction of salt from water), so this often only happens in rich countries with hot climates (like in Australia).

Rainwater Harvesting is the collection and storage of rainwater in small collection vessels or in massive reservoirs.

Image 58: Lake Mead, Hoover Dam, USA

176

3.6 WATER

Water Use

... proportions of water used for agriculture, domestic and industrial purposes in countries at different levels of economic development

Agriculture needs water to grow crops. In dry countries, this is done by irrigation. Without rainfall, water supplies need to be used for crop irrigation.

Domestic Use. In MEDCs like the UK, people use lots of water every day for washing and cooking. In some LEDCs, this isn't possible due to the scarcity or unaffordability of water.

Industrial Use. Many industries use large amounts of water in processing (like when making paper) or cooling (like in power stations).

The proportions of these 3 demands vary between each area.

- Many **LEDCs** do lots of **agriculture** – meaning that the proportion of water used for agriculture is greater in LEDCs
- In **NICs and MEDCs**, **industry** uses lots of water – meaning that the proportion of water used for industry is greater in NICs and MEDCs
- In **MEDCs**, we can afford to use massive amounts of water for **domestic** purposes – meaning that the proportion of water used for domestic use is greater in MEDCs

So, as areas become more economically developed, the demand for water, for domestic and industrial use, increases.

Image 59: Pivot Irrigation in a Cotton Farm

THEME 3
ECONOMIC DEVELOPMENT

Surplus and Deficit

Different areas have a **water deficit** or a **water surplus**. As areas become more economically developed, the demand for water also increases. The differences in supply and demand in different areas means water needs to be transferred.

Factors that determine whether there will be a water surplus or deficit:

Factor	Effect
Temperatures or the amount Evaporation	The higher the temperature, the more rainwater evaporates before there is a chance to use it. So, the higher the temperature, the likelier it is for there to be a water deficit.
Importance of Agriculture	Irrigation uses water. This means that in farming areas it is more likely for there to be a water deficit.
Amount of Precipitation	If there is little rain each year, there is a higher chance of there being a water deficit – especially if the country relies on surface water sources
Rivers or Location of Rainfall	Rivers transport water to dry areas from wet areas. If it rains a lot in a place, or if a major river flows through it, it is likelier for there to be a water surplus.
Aquifers	Groundwater is an essential source of water in dry areas. So, if aquifers are present, it is likelier for there to be a water surplus.
Population	If there are big urban areas in the region, more water will be used for domestic use – meaning that it is likelier for there to be a water deficit.
Economic Development	In MEDCs, lots of water is used for industry, power generation and domestic use – meaning that it is likelier for there to be a water deficit in dry MEDCs.

Water Shortages and Management

> Explain why there are water shortages in some areas and demonstrate that careful management is required to ensure future supplies

> The impact of lack of access to clean water on local people and the potential for economic development

Why are there Water Shortages?

Rainfall isn't equal around the world, meaning that water isn't evenly distributed. Countries with less rain may have a water deficit, creating drought. Water shortages are a result of rising demand and falling supply. The rising demand is created by economic development. Water shortages can occur if the water is

3.6 WATER

mismanaged (which is when water sources are overused), the supply unable to meet the demand.

Growing Demand:

- **Population Growth** – As the population grows, our resources are put under increasing pressure
- **Domestic Demand** – As people become wealthier, they want more baths, showers and property irrigation. These all use water
- **Agriculture** – With a growing population and migration of people, the demand for food will keep rising. Food needs water to grow, putting strain on water supplies
- **Industry** – As the population grows, and people become richer, the demand for products increases. Many industrial products (like paper) use large amounts of water in their production
- **Energy Production** – Cooling in power stations and for HEP power plants. This puts further strain on water supplies as people demand power for their technology

Falling Supply:

- **Climate Change** – The icecaps (a crucial source of fresh water) are melting, meaning the sea levels are rising. This increases the amount of salt in groundwater when sea water leaches into the ground during floods. Increased temperatures also increase **evapotranspiration**, depleting surface water stores
- **Changing Global Weather Patterns** mean that some countries and areas receive irregular supplies of rainfall, or previously plentiful supplies are no longer available
- **Ground Water Depletion** – This is because of exploitation, mismanagement and increased water extraction from wells and boreholes
- **Sewerage / Pollution** – Squatter settlements release sewerage into rivers. Also, fertilisers used in agriculture can contaminate water supplies in the same way pollutants leaching from industries do
- **Political** – Countries conflicting over water control can restrict water access in countries located downstream

Image 60: Boat in the Aral Sea

The result of the rising demand and falling supply causes the depletion of water sources, leading to water shortages.

The Impact of Water Shortages
On...

People:

- The body needs water, or it becomes **dehydrated**
    - This is due to a lack of water
    - People are **unable to work** and earn money, as water shortages make people weaker and more susceptible to disease
- **Crop Failure**
    - Could lead to **starvation**
- Water is used for domestic use in MEDCs
    - People might not be able to clean themselves, leading to **poor sanitation**

Economic Development:

- Water is required for industry
    - **Industries can't expand** if there is a water shortage
    - If industries can't expand, people could lose their jobs and their income
- If people are ill from dehydration, they **won't be able to work** and earn money affecting the country's economy
- If there is crop failure from a lack of water, people won't be able to sell their crops and earn a profit
- Because meat production demands high quantities of water, a water shortage would directly hinder the success of these industries

Managing Water
- **Water Management** – aims to reduce water waste
- **Building Dams** – to collect and store water
    - Could be an emergency water supply in water shortages
- **Limiting Water Extraction from Wells** – prevents ground water depletion
- **Rationing / Restricting water usage** – e.g. using water meters or penalties
- **Importing Water** – to support a falling supply
- **Desalination Plants** – to provide potentially limitless water from the sea
- **Drilling More Boreholes** – to use unexploited aquifers
- **Afforestation** – to increase water interception
- **Recycling unused water** – to reduce water wastage

3.6 WATER

Case Study – Lesotho
Water supply in a country or area

The Lesotho Highlands Water Project is the largest civil engineering project in Africa, and is the world's second biggest water-transfer project.

The project has 2 main objectives:

Figure 18: Map of Water supply in Lesotho

1. To create a stable and reliable water supply for Lesotho (this is being done by directing water into rivers to transport water to where it is needed most)
2. To sell surplus water to South Africa when the Johannesburg area has a water deficit. The money earnt would be fed back into Lesotho's economy, benefitting the country

This project was commissioned by the South African and Lesothan governments to develop sustainable and reliable water sources for both countries and to help to resolve issues surrounding water shortages.

181

THEME 3
ECONOMIC DEVELOPMENT

This is done by diverting about half of the Sequa or Orange river's water through dams in Lesotho. After Lesotho has taken the water it needs for its own use, the remainder is sold to South Africa, sent via the River Vaal towards Johannesburg.

Advantages for Lesotho	Disadvantages for Lesotho
<ul><li>Provides a source of funding for the Lesotho Government (this could be reinvested in other projects to improve citizens' quality of life)</li><li>Can provide water for the capital (Maseru) if there are water shortages</li><li>The Muela HEP plant generates power for Lesotho</li><li>Lots of roads and other infrastructure were built during the construction phase of the project, allowing rural villages to communicate and travel more easily</li><li>The construction of the dams created temporary jobs</li></ul>	<ul><li>Flooding of the river valleys destroyed the habitats of animals and plants</li><li>There were conflicts over land ownership. This affected the lives of rural people – threatening both their farming lifestyle and their villages</li><li>People who lived in the flooded valleys were displaced</li></ul>

Overall, the project has a positive impact, generating money from selling power and water to South Africa, and by linking remote communities from the construction of new roads and infrastructure.

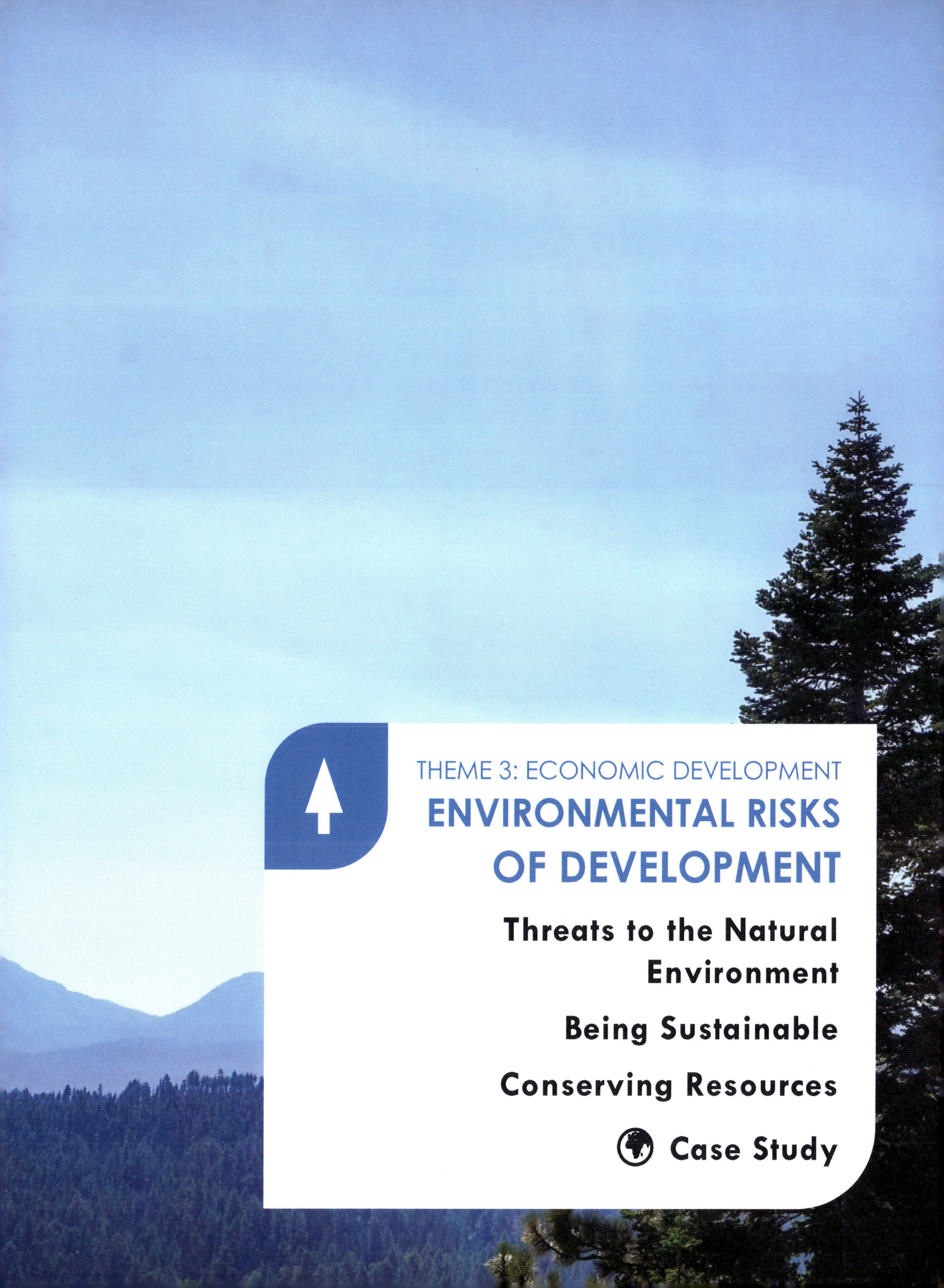

THEME 3: ECONOMIC DEVELOPMENT

ENVIRONMENTAL RISKS OF DEVELOPMENT

Threats to the Natural Environment

Being Sustainable

Conserving Resources

 Case Study

THEME 3
ECONOMIC DEVELOPMENT

Describe how economic activities may pose threats to the natural environment and people, locally and globally

Demonstrate the need for sustainable development and management

Understand the importance of resource conservation

Threats to the Natural Environment

Describe how economic activities may pose threats to the natural environment and people, locally and globally

Threats to the natural environment (including soil erosion, desertification, enhanced global warming and pollution [water, air, noise, visual])

Why is there a threat?
Since the industrial revolution (from around 1800) there has been:

- Coal mining and burning of coal (increasing the amounts of greenhouse gases, and producing air pollution)
- The expansion of towns and cities – from urbanisation when people seek work in the new factories, increasing light pollution
- The creation of new transport routes to link the growing industries. This caused noise pollution and increased the amount of greenhouse gases emitted
- The destruction of natural vegetation to provide land for farming
- Deforestation, meaning habitats were destroyed and carbon capture reduced
- Exploitation of soils – stripping soil of its nutrients by over-farming agricultural crops
- Washing of fertilisers from increased food production into water supplies
- Water pollution from increased industry – contamination of water sources by toxic chemicals

Soil Erosion
This is caused by either **wind blowing away soil** or by **water running down slopes**.

<u>By the wind:</u>
- There needs to be little precipitation, making the soil dry and loose
- Strong winds are needed to blow away the soil

<u>By running water:</u>
- Slopes must be steep for the water to run down them
- The rainfall must be heavy enough for all of it to soak into the ground, saturating the soil. This allows surface run-off, which removes soil

3.7 ENVIRONMENTAL RISKS OF ECONOMIC DEVELOPMENT

- Water flows in sheets of water on gentle slopes, and in channels on steep slopes

Erosion is the result of **soil being exposed**, for instance, by not being covered by vegetation. Another factor that causes soil erosion is the soil being loose, damaged by poor agricultural practice – meaning it loses its structure.

Humus is decayed vegetation. It helps soil particles to stick together. The reduction in humus causes the soil structure to break down – making soil erosion more likely.

Overall, soil erosion occurs as a result of poor agricultural practice – like over grazing or over cultivating.

Soil conservation is the process of reducing soil erosion:

- **Terracing** – retaining walls built on a slope, preventing the water from running down the slope and removing soil. Water becomes trapped, soaking into the ground, preventing soil erosion from running water
- **Crop Rotation** – a different crop is grown on a plot of land each year for a few years before the 1st crop can be grown again. Different crops take different nutrients from the soil, so crop rotation prevents soil from becoming exhausted. Exhausted soil leads to loose soil structure that increases the chance of erosion
- **Irrigation** – keeps the soil moist, preventing it from being picked up by the wind and reducing the chance of it being carried away by running water
- **Afforestation** – trees stop strong winds and heavy rainfall from reaching the soil. Roots also hold the soil together, giving the soil structure
- **Wind Breaks** – rows of trees are grown at a 90-degree angle to the area's prevailing winds. This reduces the speed of the wind, meaning it doesn't have enough energy to pick up and erode the soil.

Image 61: Terraced Rice Fields, Yunnan Province, China

THEME 3
ECONOMIC DEVELOPMENT

Image 62: Result of Running Water Soil Erosion - Desertification

Desertification

Occurs when there is lots of soil erosion, dry soil and lack of humus.

Desertification is when land increasingly becomes like a desert.

But why does it happen? These are some factors linked to economic development that could help to cause desertification:

- **Overgrazing removes grass and vegetation, exposing the soil**. When there are no roots to provide soil structure, it is easier for the soil to be blown away and eroded. This soil erosion could lead to desertification. Overgrazing could also could lead to **less humus** in the soil, as the grazing strips the ground of vegetation – meaning less humus is formed. This means that the soil structure will break down, crumbling into individual particles that are lighter and more likely to be blown away by the wind. The purpose of overgrazing is to exploit the land for maximum profit, and to feed the growing population.
- **Population growth** leads to more **vegetation being removed for firewood** in LEDCs. This would mean that the soil is more exposed, and consequently more liable to soil erosion by the wind or running water. Another impact of population growth is that the **water table could fall**, as the population exploits the groundwater and aquifers through boreholes to supply water for increased industry, agriculture and domestic use. This would dry out the soil, meaning it is more likely for soil erosion to occur because the soil becomes lighter and more fragile.
- **Overcultivation** *is* a result of population growth – because population growth increases the demand for food. This would lead to **soil exhaustion**, meaning that nutrients in the soil are depleted, increasing the chance of soil erosion (because soil structure is lost).
- **Drought**, as a result of enhanced global warming, could make soil erosion more likely. Dry soil is lighter, so can more easily be eroded by wind and running water.

3.7 ENVIRONMENTAL RISKS OF ECONOMIC DEVELOPMENT

Enhanced Global Warming

The Earth is continually warmed by the sun's radiation. At night, the earth loses heat through the atmosphere. The Earth's temperature remains constant when the incoming and outgoing infrared radiation are equal.

Greenhouse gases in the atmosphere prevent infrared radiation from escaping through the atmosphere. They keep the Earth at the ideal temperature for life when in constant, natural concentrations.

However, recent human activity has meant that the amount of greenhouse gases in the atmosphere has increased. This means that heat is prevented from escaping out into space. Consequently, world temperatures have risen by around half a degree in the past century. Some estimates suggest that a further rise of around 2 degrees could occur by 2100. This process, where world temperatures rise, is called **enhanced global warming**.

Why is there enhanced global warming?

Firstly, it is useful to note that one of the major contributors to global warming is **carbon dioxide gas**.

- **Burning fossil fuels** for energy releases **carbon dioxide**
    - Because MEDCs and NICs account for most of the world's energy use, they are mainly responsible for the increased carbon dioxide emissions
    - Industrialisation increases carbon dioxide emissions because factories burn fossil fuels in furnaces
- **Methane** – released from decaying organic matter
    - Methane emissions have increased due to increased demand for meat, and consequently, increased farming of cattle, which produces more methane gas
- **Nitrous oxide** – another greenhouse gas, released from cars, power plants and some fertilisers
    - Increased food production for growing populations means the use of fertilisers, and emissions of nitrous oxide, have increased
    - More people driving in cars increases emissions of nitrous oxide
    - More people increases the demand for power, meaning that more nitrous oxide is emitted from power stations
- **CFCs** - released from aerosols and fridges
    - This depletes ozone in the atmosphere, meaning the Earth is less insulated from the Sun's radiation, so receives more infrared radiation. Unfortunately, the damage has already been done, and they now shouldn't be used at all, including in developing countries

What are the effects of enhanced global warming?

- **Sea temperatures will rise** – as the sea warms, the water will expand – meaning that the **sea level will rise**
- **Ice caps / Glaciers will melt** – due to increased air temperatures. The release of this water into the sea could further raise the sea level. This could **flood** parts of Bangladesh and totally submerge many islands, **displacing people** and **destroying coastal habitats**
- **Weather patterns will change** – some places will become wetter, and others drier (meaning drought may become more common in some areas, like in the Horn of Africa)

What can we do to reduce enhanced global warming?

- **Using Renewable Energy and Nuclear Power** – These alternative forms of energy production don't release greenhouse gases, so won't contribute to enhanced global warming
- **Reducing Deforestation** – This means that the trees capturing carbon dioxide will be retained, meaning that more carbon dioxide will be taken out of the atmosphere, reducing enhanced global warming
- **Improving Energy Efficiency** – This would mean that less carbon dioxide is emitted for each unit of energy – meaning that overall, less carbon dioxide is emitted into the atmosphere
- **Reducing Industry's Emissions** – This would reduce greenhouse gas emissions from industry, reducing the enhanced global warming effect
- **Practicing Sustainable Living**
- **Using Sustainable Transport** – Alternative forms of transport could release no carbon dioxide, so won't contribute to enhanced global warming
- **Carbon capture to reduce CO_2 Emissions**

Image 63: Smoke from Industry

3.7 ENVIRONMENTAL RISKS OF ECONOMIC DEVELOPMENT

Pollution

There are 4 types of pollution:

- **Air** – Contamination of the atmosphere by emissions – e.g. by carbon dioxide, nitrous oxide and particulates
- **Noise** – Noise from industry or vehicles – e.g. from quarries or factories
- **Water** – Contamination of a water supply – e.g. by industry releasing toxic waste into rivers
- **Visual** – Impacting the beauty and appearance of the natural world – e.g. quarries – or something that spoils the beauty of the landscape

Image 64: Quarry, Hühnerberg

Being Sustainable

Demonstrate the need for sustainable development and management

Sustainability is fulfilling our resource requirements without hindering the access to these benefits for future generations. This means that we should only use the Earth's resources at the rate that they can be replenished.

Goals of sustainability:

- To maintain and improve natural beauty and wildlife that resides in nature
- To ensure the public understands the importance of the environment
- To ensure that economic development can occur without impacting the environment

By developing in a sustainable manner, we can reduce our impact on the environment. This would allow us to preserve the Earth, its life, resources and beauty for future generations.

By managing development, we can achieve our goal of reducing our environmental impact.

THEME 3
ECONOMIC DEVELOPMENT

Image 65: Pieniny National Park, Poland

Ways of managing economic development to care for the environment:

- **Creating national parks** to protect areas of environmental significance
- **Enacting legislation** to reduce emissions from cars and industry, limiting the amount of greenhouse gases emitted – reducing the enhanced global warming effect
- **Reducing landfill sites** to reduce methane emissions created by decaying organic matter
- **Protecting areas of forest**, to reduce the likelihood of their deforestation

3.7 ENVIRONMENTAL RISKS OF ECONOMIC DEVELOPMENT

- **Educating people** about the importance of conserving resources and caring for the environment
- **Educating farmers** about how to reduce soil erosion, improve sustainability of farming and to reduce the chance of desertification
- **Managing urban growth**, to reduce the destruction of greenfield sites
- Controlling the **locations** that industry can occur in, ensuring they are far from wildlife habitats that could be destroyed by industrial activity
- **Developing infrastructure** and **managing pollution** so the urban environment would be suitable for future generations
- **Conserving resources** for future generations
- Ensuring countries meet to establish achievable international **environmental policies and goals**
- Ensuring that economic activities won't reduce the **availability of Earth's resources** for future generations

Resource Conservation

Understand the importance of resource conservation

Conservation is protecting the natural environment and its resources for the future. These could be natural resources (e.g. precious metals, energy or climate) or living resources (e.g. animals or habitats). This is to prevent resource depletion (to improve sustainability) and to prevent extinction.

How does reducing, reusing and recycling help us to conserve natural resources?

- Resources last longer
- Energy is saved, because recycling means we don't have to use energy to extract new raw materials from the ground
- We can use previously extracted resources again
- We can make use of waste products

Why is it important to conserve natural resources and the environment?

- By conserving them, we can **meet the future demands for products**
- Many **resources are finite**
- Life wouldn't be possible without **clean air and water**
- To **protect habitats**
- To **prevent global warming**
- To maintain **balance** in ecosystems
- To **avoid conflict** over abuse of resources
- Soil needs to have **nutrients** for plants and crops to grow
- Forests act as **carbon sinks** for our carbon dioxide emissions, meaning we can **reduce further enhanced global warming**

THEME 3
ECONOMIC DEVELOPMENT

Case Study – The Aral Sea, Kazakhstan and Uzbekistan
An area where economic development is taking place causing the environment to be at risk

Before 1965, the Aral Sea occupied a massive area in Kazakhstan and Uzbekistan. It was the world's 4th largest lake, fed by 2 of Asia's major rivers.

The climate around the Aral Sea is extremely arid, with very high summer temperatures of up to 50 degrees. The winters here are also very cold and there is low precipitation throughout the year.

Image 66: The Aral Sea in 1989 and 2008

3.7 ENVIRONMENTAL RISKS OF ECONOMIC DEVELOPMENT

This means that agriculture is much more difficult, meaning people made a living by fishing in the sea. Consequently, people in the area were employed as fishermen and in supporting industries on the lake's shore.

In the 1960s, Kazakhstan and Uzbekistan were still part of the Soviet Union. The government wanted to use the water in the Amu Darya (which feeds the Aral Sea) to grow crops (wheat and cotton). To grow these crops, huge canals, for irrigation, were built to divert river water into fields. So, 3 million hectares of crop land was created.

However, because the rivers were diverted away from the Aral Sea, less water reached the sea, meaning that it began to rapidly shrink in size.

Effects of the shrinking Aral Sea (mismanaged economic activity):

Effects on People	Effects on Environment
<ul><li>Fishing is much harder – so people lost their jobs and their source of income</li><li>Water is now further from the towns, causing water shortages</li><li>Ships were abandoned in the dried-up lake</li><li>Fish died, meaning fishing industries collapsed</li><li>Drinking water was polluted by fertilisers used in farming</li><li>Violent dust storms blew salt, sand and chemicals into the air – meaning people developed breathing diseases like Tuberculosis</li><li>People became protein-deficient as they could no longer eat fish. This malnutrition created a high infant mortality rate</li><li>Summers grew hotter, and winters colder – as the lake could no longer regulate the local climate</li></ul>	<ul><li>Salinity in the lake increased as the water level fell – meaning fish died because they couldn't cope with the suddenly very salty water</li><li>Fertilisers and pesticides used in the intensive farming of cotton and wheat polluted water that was used for irrigation and for drinking</li><li>Summers became hotter and winters colder – as the lake could no longer regulate the local climate</li></ul>

THEME 3
ECONOMIC DEVELOPMENT

Image 67: Abandoned ship in the Aral Sea, Near Aral, Kazakhstan

Useful fact: The **Amu Dayra** and the **Syr Dayra** fed the Aral Sea

MEDIA LISENCES

LICENCES FOR MEDIA USED

CC BY 4.0 https://creativecommons.org/licenses/by/4.0/

CC BY SA 4.0 https://creativecommons.org/licenses/by-sa/4.0/

CC BY 3.0 https://creativecommons.org/licenses/by/3.0/

CC BY-SA 3.0 https://creativecommons.org/licenses/by-sa/3.0/

CC BY 2.5 https://creativecommons.org/licenses/by/2.5/

CC BY-SA 2.5 https://creativecommons.org/licenses/by-sa/2.5/

CC BY 2.0 https://creativecommons.org/licenses/by/2.0/

CC BY-SA 2.0 https://creativecommons.org/licenses/by-sa/2.0/

CC BY-SA 1.0 https://creativecommons.org/licenses/by-sa/1.0/

FAL http://artlibre.org/licence/lal/en/

INFORMATION

ACKNOWLEDGMENTS

Images

Image 1: For a prosperous, powerful nation and a happy family, please practice family planning - by Venus - China One Child Policy, CC BY 2.0, https://commons.wikimedia.org/w/index.php?curid=2964631 13

Image 2: Flooding in Bangladesh – Public Domain, https://commons.wikimedia.org/w/index.php?curid=1701157 14

Image 3: Australian Outback - by Andy Mitchell from Glasgow, UK - CC BY-SA 2.0, https://commons.wikimedia.org/w/index.php?curid=27178433 ...15

Image 4: The Crowded Streets of Dhaka - by Soman assumed (based on copyright claims), Own work assumed (based on copyright claims), CC BY 2.5, https://commons.wikimedia.org/w/index.php 16

Image 5: Refugees arriving at Lesvos island - by Ggia - Own work, CC BY-SA 4.0, https://commons.wikimedia.org/w/index.php?curid=45303280 ...20

Image 6: Crowds on George Street, Sydney, Australia - by Mike Weber, CC SA-BY 2.0, https://commons.wikimedia.org/wiki/File:Crowds_on_George_Street_(10098460084).jpg 22

Image 7: Ochanomizu, Tokyo, Japan - by Kabelleger / David Gubler - Own work: https://bahnbilder.ch/picture/25855, CC BY-SA 4.0, https://commons.wikimedia.org/w/index.php?curid=60325982 ...26

Image 8: Aircraft at Manchester Airport Terminal 2 - by profile500, Public Domain, https://commons.wikimedia.org/wiki/File:Aircraft_at_Manchester_Airport_Terminal_2.jpg 39

Image 9: Smog in Cairo - by Sturm58 at English Wikipedia - Transferred from en.wikipedia to Commons., CC BY-SA 3.0, https://commons.wikimedia.org/w/index.php?curid=2823593 43

Image 10: The Printworks, Manchester – by David Dixon, CC BY-SA 2.0, geograph.org.uk/p/1913740 47

Image 11: Urban Sprawl in South Los Angeles - by Alfred Twu, Public Domain CC0 1.0, https://commons.wikimedia.org/wiki/File:South-Los-Angeles-110-and-105-freeways-Aerial-view-from-north-August-2014.jpg 49

Image 12: Traffic Congestion on the I-110, Los Angeles, CA – by Coolcaesar, CC BY 4.0, https://commons.wikimedia.org/wiki/File:Harborfreeway2.jpg 50

Image 13: Houses in an Informal Settlement - by SuSanA Secretariat - Houses in informal settlement Uploaded by Elitre, CC BY 2.0, https://commons.wikimedia.org/w/index.php?curid=22163467 ...51

Image 14: Dharavi (near Mahim Junction), Mumbai, India - by A.Savin (Wikimedia Commons · WikiPhotoSpace) - Own work, FAL, https://commons.wikimedia.org/w/index.php?curid=48098620 ...55

Image 15: Arenal Volcano- by Matthew.landry at English Wikipedia, CC BY-SA 2.5, https://commons.wikimedia.org/w/index.php?curid=4398466 62

Image 16: Mount St. Helens Erupting, USA - Public Domain, https://commons.wikimedia.org/w/index.php?curid=260607 65

Image 17: A Volcanologist on Etna - by Laurenti andré - Own work, CC BY-SA 3.0, https://commons.wikimedia.org/w/index.php?curid=31181825 ...67

Image 18: Fissure on Fimmvörðuháls (part of the Eyjafjallajökull system)) - by Boaworm - Own work, CC BY 3.0, https://commons.wikimedia.org/w/index.php?curid=10026499 ...68

Image 19: Kobe Port Memorial Park – Uploaded by S lawojar - CC BY-SA 3.0, https://commons.wikimedia.org/w/index.php?curid=43961 73

Image 20: The Hydrological Cycle - by Ehud Tal - Own work, CC BY-SA 4.0, https://commons.wikimedia.org/w/index.php?curid=47658638 ...76

Image 21: - Mississippi River South of Muscatine, Iowa, USA - by Ken Lund from Reno, Nevada, USA, CC BY 2.0, https://commons.wikimedia.org/w/index.php?curid=45408631 ...83

Image 22: Lena River Delta False Colour - (Landsat) - Public domain - via Wikimedia Commons - https://upload.wikimedia.org/wikipedia/commons/f/fb/Lena_River_Delta_-_Landsat_2000.jpg 84

Image 23: Hoover Dam, Arizona, USA - by Marionzetta - http://www.flickr.com/photos/marionzetta/2904641633/, GFDL, https://commons.wikimedia.org/w/index.php?curid=10361658 ...87

Image 24: Mekong River at Luang Prabang, Laos - by Allie Caulfield - http://www.flickr.com/photos/wm_archiv/3923953618/in/set-72157622393374604/, CC BY 2.0, https://commons.wikimedia.org/w/index.php?curid=8570339 89

Image 25: Cliffs of Bonifacio - by Myrabella - Wikimedia Commons, CC BY-SA 3.0, https://commons.wikimedia.org/w/index.php?curid=28110274 ...95

Image 26: Wavecut Platform- by Ian Balcombe, CC BY 2.0, https://commons.wikimedia.org/w/index.php?curid=13796437 ...95

Image 27: Bruny Island Spit- by JJ Harrison (jjharrison89@facebook.com) - Own work, CC BY-SA 2.5, https://commons.wikimedia.org/w/index.php?curid=5792158 98

Image 28: Mangrove Forest, Puerto Rico - by Boricuaeddie - Own work, CC BY-SA 3.0, https://commons.wikimedia.org/w/index.php?curid=3055361... 101

Image 29: Groynes at Hurst Castle on Hurst Spit - by JamesBrownNFNPA - Own work, CC BY-SA 3.0, https://commons.wikimedia.org/w/index.php?curid=7583616... 104

Image 30: Rain Gauge - by Bocholter - Self-photographed, Public Domain, https://commons.wikimedia.org/w/index.php?curid=8250104... 106

Image 31: Six's Thermometer - by Lumos3 at the English language Wikipedia, CC BY-SA 3.0, https://commons.wikimedia.org/w/index.php?curid=3181381... 107

Image 32: Dry and Wet Bulb Thermometers - by Internet Archive Book Images - https://www.flickr.com/photos/internetarchivebookimages/14764108305/Source book page: https://archive.org/stream/journalofroyalho4187roya/journalofroyalho4187roya#page/n38/mode/1up, No R................................. 107

Image 33: Sunshine Recorder, Botanical Garden, Funchal, Madeira - by Stefan Bellini - Own work, CC0, https://commons.wikimedia.org/w/index.php?curid=49921462 108

Image 34: Modern Aneroid Barometer - by Agnellous, Own work, Public Domain, https://commons.wikimedia.org/w/index.php?curid=1268977... 109

Image 35: Barograph – uploaded by Saperaud~commonswiki, CC BY-SA 3.0, https://commons.wikimedia.org/w/index.php?curid=295958..... 109

Image 36: Anemometer - derived (picture edited) by Staycoolandbegood; original by Nordelch - CC BY-SA 3.0, https://commons.wikimedia.org/w/index.php?curid=402480..... 110

Image 37: Wind Vane, Farnborough - by KyleH, CC BY 3.0, https://commons.wikimedia.org/w/index.php?curid=54244625 110

Image 38: Stevenson Screen - by Bidgee - Own work, CC BY 3.0, https://commons.wikimedia.org/w/index.php?curid=5541848... 113

Image 39: Rainforest in Belize - by Cephas - Own work, CC BY-SA 4.0, https://commons.wikimedia.org/w/index.php?curid=48674234 118

Image 40: Dunes in The Sahara Desert – uploaded by Pir6mon, CC BY-SA 3.0, https://commons.wikimedia.org/w/index.php?curid=1206803... 119

Image 41: Rainforest at Puentes Colgantes, near Arenal Volcano, Costa Rica - by Hans Hillewaert /, CC BY-SA 3.0, https://commons.wikimedia.org/w/index.php?curid=10203677 122

Image 42: Arizona Desert - by Sam Mugraby, Photos8.com, CC BY 2.0, https://commons.wikimedia.org/w/index.php?curid=8345847... 125

Image 43: Camel in Desert - by Sam Mugraby, Photos8.com, CC BY 2.0,

196

AKNOWLEDGEMENTS

Image 44: Charcoal Production and Deforestation - by User Kelberul on de.wikipedia - Originally from de.wikipedia; description page is (was) here16:05, 17. Sep 2004 Kelberul 2048 x 1536 (351.098 Byte) (Holzkohlegewinnung im Regenwald - GNU-FDL - selbst fotografi... 129

Image 45: Male Bornean Orangutan - by Eric Kilby from Somerville, MA, USA - Male Bornean Orangutan - Big Cheeks, CC BY-SA 2.0, https://commons.wikimedia.org/w/index.php?curid=40995032. 130

Image 46: Dunes in the Namib Desert - by Robur.q - Own work, CC BY-SA 4.0, https://commons.wikimedia.org/w/index.php?curid=42942814. 133

Image 47: FCAB Freight Train, Bolivia - by Diego Delso, CC BY-SA 4.0, https://commons.wikimedia.org/w/index.php?curid=47641444. 139

Image 48: Nike's Headquarters, OR, USA - by Carson, Brandon - Flickr, CC BY 2.0, https://commons.wikimedia.org/w/index.php?curid=52974841. 142

Image 49: A Corn Farm, Iowa, USA - by inkknife_2000 (7.5 million views +) - https://www.flickr.com/photos/23155134@N06/15277889101/, CC BY-SA 2.0, https://commons.wikimedia.org/w/index.php?curid=57522665. 146

Image 50: Rice Farm, Bangladesh - by Raiyan - Own work, CC BY 3.0, https://commons.wikimedia.org/w/index.php?curid=5397858 ... 149

Image 51: Transport is an Important Factor when locating industry - FCCA Freight Train, Peru - by Kabelleger / David Gubler (http://www.bahnbilder.ch) - Own work: http://www.bahnbilder.ch/picture/13011, CC BY-SA 3.0, https://commons.wikimedia.org/w/index.php?cur....................... 153

Image 52: Silicon Valley, USA - by Anthonyavalos408 - Own work, CC BY-SA 4.0, https://commons.wikimedia.org/w/index.php?curid=51286408. 155

Image 53: Toyota Factory, Burnaston - by Nick Moyes - Own work, CC BY-SA 4.0, https://commons.wikimedia.org/w/index.php?curid=57262604. 156

Image 54: by Jesus Rodrigo Jara, CC BY 3.0, https://commons.wikimedia.org/w/index.php?curid=12473229. 164

Image 55: People often have to travel long distances to gather fuelwood - by Stephen Codrington, CC BY 2.5, https://commons.wikimedia.org/w/index.php?curid=221260 172

Image 56: Nuclear power plant, Cattenom, France - by Stefan Kühn - Own work, CC BY-SA 3.0, https://commons.wikimedia.org/w/index.php?curid=94202 172

Image 57: HGPS, Iceland - http://www.flickr.com/photos/thinkgeoenergy/4473298115/, CC BY 2.0, https://commons.wikimedia.org/w/index.php?curid=10907139. 174

Image 58: Lake Mead, Hoover Dam, USA - Public Domain, https://commons.wikimedia.org/w/index.php?curid=12001684. 176

Image 59: Pivot Irrigation in a Cotton Farm - Public Domain, https://commons.wikimedia.org/w/index.php?curid=4551861 ... 177

Image 60: Boat in the Aral Sea – Public Domain, https://commons.wikimedia.org/w/index.php?curid=658367 179

Image 61: Terraced Rice Fields, Yunnan Province, China - by Jialiang Gao, www.peace-on-earth.org - Original Photograph, CC BY-SA 3.0, https://commons.wikimedia.org/w/index.php?curid=2926830 ... 185

Image 62: Result of Running Water Soil Erosion - Desertification - by Chris English, CC BY-SA 3.0, https://commons.wikimedia.org/w/index.php?curid=56575639. 186

Image 63: Smoke from Industry – by Uwe Hermann, CC BY-SA 2.5, https://commons.wikimedia.org/w/index.php?curid=902877/index.php?curid=902877work, CC BY-SA 3.0, https://commons.wikimedia.org/w/index.php?curid=8735834 ... 188

Image 64: Quarry, Hühnerberg - by Wolkenkratzer - Own work, CC BY-SA 3.0, https://commons.wikimedia.org/w/index.php?curid=19911603. 189

Image 65: Pieniny National Park, Poland - by Jerzy Opioła - Own work, CC BY 2.5, https://commons.wikimedia.org/w/index.php?curid=1306151 ... 190

Image 66: The Aral Sea in 1989 and 2008 - by NASA, derivative work by Zafiroblue05 at en.wikipedia - 1989: earthobservatory.nasa.gov, aral sea 1989 250mFile:Aral Sea 05 October 2008.jpg, Public Domain, https://commons.wikimedia.org/w/index.php?curid=5957380 ... 192

Image 67: Abandoned ship in the Aral Sea, Near Aral, Kazakhstan - Public Domain, https://commons.wikimedia.org/w/index.php?curid=12751640. 194

Figures

Figure 1: Diagram of the Demographic Transition Model (DTM) - by Max Roser - CC BY-SA 4.0, https://commons.wikimedia.org/w/index.php?curid=47036003 Source: http://www.OurWorldInData.org/data/population-growth-vital-statistics/world-population-growth 10

Figure 2: Kenya's Population Structure in 2016 - CIA World Factbook – Public Domain - https://www.cia.gov/library/publications/the-world-factbook/geos/ke.html 24

Figure 3: Japan's Population Structure in 2016 2016 - CIA World Factbook – Public Domain - https://www.cia.gov/library/publications/the-world-factbook/geos/ja.html 25

Figure 4: Position of Tectonic plates around the world - Public Domain, https://commons.wikimedia.org/w/index.php?curid=35511 58

Figure 5: Conservative Plate Boundary - by domdomegg - Own work, CC BY 4.0, https://commons.wikimedia.org/w/index.php?curid=50782243 ... 60

Figure 6: Destructive Plate Boundary - by domdomegg - Own work, CC BY 4.0, https://commons.wikimedia.org/w/index.php?curid=45732560 ... 60

Figure 7: Constructive plate boundary - by domdomegg - Own work, CC BY 4.0, https://commons.wikimedia.org/w/index.php?curid=50772217 ... 60

Figure 8: Structure of a Stratovolcano (Left) – by William Crochot - File:Spaccato vulcano.png, CC BY-SA 4.0, https://commons.wikimedia.org/w/index.php?curid=37372932 ... 64

Figure 9: Waterfall Formation - by Cradel - Derived from Waterfall formation23.png, CC BY-SA 3.0, https://commons.wikimedia.org/w/index.php?curid=5960898 81

Figure 10: Longshore Drift - by Yefi - Own work, Public Domain, https://commons.wikimedia.org/w/index.php?curid=7349541 94

Figure 11: Public Domain, https://commons.wikimedia.org/w/index.php?curid=425700 100

Figure 12: Mercury Barometer - Public Domain, https://commons.wikimedia.org/w/index.php?curid=2843015 ... 109

Figure 13: Cloud Cover Symbols - by Frasmacon - Own work, CC BY-SA 3.0, https://commons.wikimedia.org/w/index.php?curid=25497628. 111

Figure 14: Cloud Types - by Valentin de Bruyn / Coton - Own work, CC BY-SA 3.0, https://commons.wikimedia.org/w/index.php?curid=17899555. 111

Figure 15: World Distribution of Tropical Rainforests - Public Domain, https://commons.wikimedia.org/w/index.php?curid=1550301 ... 117

Figure 16: Rain shadow effect - by domdomegg - Own work, CC BY 4.0, https://commons.wikimedia.org/w/index.php?curid=45877009. 131

Figure 17: Horn of Africa Location - https://commons.wikimedia.org/w/index.php?curid=2992371 ... 150

Figure 18: Map of Water supply in Lesotho - by Tambo, CC BY-SA 3.0, https://commons.wikimedia.org/w/index.php?curid=18085250. 181

197

INFORMATION

COMMAND WORDS

Word	Meaning
Define / State the meaning of / What is meant by	Give the definition or meaning of the word or phrase you are given
Describe	Write what something is like or where it is. Illustrate with words the aspect you are told to describe
Draw	Make a sketch of the thing you are told to draw
Explain / Account for / Give Reasons for	Write why something happens
Give your views / Comment on	Give your opinion
How far do you agree	Use specific evidence to judge a statement
Identify	Pick out a certain piece of information from what you are given (e.g. photograph / data)
Illustrate your answer	Use specific examples or diagrams to support your answer
Insert / Label	Add details to an illustration as instructed
Justify	Say why you chose something. Back up your answer / argument
List	Identify and name multiple points / features as required
Locate	Find something (/state location of) something on a provided map or diagram
Name	State / Specify / Identify. This is what the thing you are told to name is called
Predict	Use your own knowledge (if given, with provided information) to predict what might happen
Refer to / With reference to	Use aspects / ideas in the provided map / photo / diagram or case study material in your answer
State	Briefly refer to an aspect in a short statement or a single word
Study	Look carefully at / analyse the provided material referenced
Suggest (+why)	Write your ideas about something / knowledge of something (explain / state why this is so [/link it back])
To what extent	Make judgements from evidence to evaluate a statement
Use / Using the information provided	Base your answer to the question on the referenced material
With the help of information in	Write your answer using information provided **and** with your own knowledge
What differences are shown between A and B	Use comparative statements in your answer to describe the changes involved as A changes to B

Please refer to the latest syllabus information, available on the course's website, for the latest official command word definitions.

GLOSSARY

Accessibility	How easy it is to get to something
Acid Lava	Lava that contains a high amount of silica, so has a very high viscosity. It forms steep sided volcanic cones, usually in a more violent eruption. The sides are steep because acid lava flows slowly
Active Volcano	A volcano that has recently erupted
Adaptations	Changes an organism makes to become better suited to its environment
Afforestation	Planting of trees in a location where there were no trees before
Aging Population	The shift of age distribution in a population to the older side
Agriculture	The practice of farming
AIDS	Caused by HIV, it is a condition where immunity is reduced - meaning that the patient is more likely to become infected by diseases
Air Pollution	Contamination of the atmosphere by emissions
Anemometers	Measure the wind's velocity by measuring the velocity that its cups rotate when driven by wind
Annual Temperature Range	The difference between the highest and lowest temperatures in a year
Anti-Natalist Policy	Aims to reduce population growth by encouraging couples to have less children
Aquifers	Water-containing layers of rock
Arable Farms	Farms that grow crops
Arid	Being too dry to support vegetation growth
Ash Clouds	Clouds of ash emitted during a volcanic eruption
Aspect	The direction that a slope faces
Asylum Seekers	People who are trying to get refuge in a foreign country
Atmosphere	The gases or air surrounding the earth
Attrition	The process of the load being eroded, becoming rounder over time
Backwash	The motion of waves withdrawing from the coast
Barometers	Instruments used to measure pressure
Basic Lava	Lava that contains a lower amount of silica, so has a low viscosity. It forms shallow sided volcanic cones, usually in a gentler eruption. The sides are shallower because basic lava is very fluid
Bay	Areas between headlands which have been eroded away, forming a gentle curve
Beach	A coastal deposition feature formed when sand or pebbles are deposited on the coastline
Beaufort Scale	A system that sorts wind speeds into classes, categorised by damage caused
Bedload	Load found on the riverbed of a river
Biodiversity	The variety of life in an area
Biome	A large ecosystem
Birth Control	The prevention of unwanted pregnancies
Birth Rate	Number of Live Births per 1000
Bradshaw Model	A model used to show how characteristics of a river change from the source to the mouth
British Pound (£)	The currency used in the United Kingdom
Brownfield Land	Land where industry used to be located

INFORMATION

Budget Airline	An airline that has cheap flights. They make flights cheaper by removing all luxuries, like hold luggage and food, from the base fare
Buttress Roots	Very long roots that come off tall trees to gather more nutrients and to provide additional support
Capital	Money
Capital City	The most important settlement in an area, often containing central government buildings and administrative services
Carbon Neutral	When the amount of carbon dioxide taken out of the atmosphere is equal to the amount of carbon dioxide emitted into the atmosphere
Central Business District (CBD)	The centre of the city, mainly consisting of high rise buildings
Channel Flow	When water travels in channels, like rivers
City	A settlement where a large number of people live and work
Civil War	A way between people in the same country
Clear Felling	Felling large areas of forest at once
Cliffs	Vertical slabs of rocks formed by weathering and marine erosion
Climate	The average weather conditions over a period of time
Closed System	A system which is self-contained and nothing enters or leaves the system
Clouds	Condensed water vapour in the atmosphere
Coast	The place where the land meets the sea
Coastal Deposition	The process involving the sea's load being dropped or left on a beach by the waves
Coastal Management	Controlling erosion, flooding and longshore drift
Coastline	The outline of the edge of the land on a map
Collision Plate Boundaries	Where 2 continental plates collide
Commercial Farming	Farming that produces food for sale
Composite Indicator	An indicator of development that is based on multiple other indicators of development
Concordant Coastlines	Coastlines made up of the same type of rock, so have less bays and headlands
Confluence	Where 2 rivers join or meet
Congestion	Too much traffic in an area - leading to widespread accumulation of queuing vehicles
Congestion Charge	A fee that drivers have to pay to enter the city centre in some cities - to reduce congestion and pollution in city centres
Conservation	Protecting the natural environment and its resources for the future. These could be natural resources (e.g. precious metals, energy or climate) or living resources (e.g. animals or habitats). This is to prevent resource depletion (to improve sustainability) and to prevent extinction the natural environment and its resources for the future
Conservative Plate Boundaries	Where 2 plates move alongside each other
Constructive Plate Boundaries (Divergent)	Where 2 plates move apart
Constructive Waves	Waves that carry out deposition
Consumers (Ecosystem)	Organisms which feed on the producers and any consumer in a lower trophic level

GLOSSARY

	(lower down) in the food web	**Dependency Ratio**	Proportion of dependent people to the working population
Consumers (Services)	People who buy products and use services	**Dependent People**	People who rely on the working population - e.g. the elderly or the young
Continental Crust	Dense tectonic plates where most of the planet's landmasses are located	**Deposition**	When a river's (or sea's) load is left behind because the river (or sea) no longer has enough energy to carry it
Conurbation	An extended urban area formed as a result of towns being absorbed into one or more cities' suburbs		
Convection Currents	Currents in the mantle which move tectonic plates	**Depression Table**	A table used to work out the relative humidity from the wet and dry bulb temperatures in a hygrometer
Coral Polyps	Tiny marine animals that form reefs when they live in colonies		
Coral Reefs	A marine feature formed by growth of coral polyps	**Derelict**	In poor condition due to disuse
Corrasion (Abrasion)	Erosion caused by the load rubbing against a surface, scraping off material	**Desalination**	When sea water is distilled to remove the salt content so we can use and drink it
Corrosion (Solution)	The dissolving of rocks. Limestone dissolves in water, so material can be eroded from it in solution	**Desert-ification**	When land increasingly becomes like a desert
		Destination Country	The country which a person is travelling to
Crust	The outer layer of the Earth where we live. It is composed of massive Tectonic Plates	**Destructive Plate Boundaries (Convergent)**	Where less dense oceanic crust is subducted under denser continental crust
Culture	The customs, beliefs, values, and traits of a group of people		
Dams	Large walls that hold back water in reservoirs, allowing the water to be released slowly over time - allowing power to be generated	**Destructive Waves**	Waves that carry out erosion
		Dikes	A long embankment built to prevent flooding
		Discordant Coastlines	Have alternating rock strata of hard and soft rock, so have more bays and headlands
Death Rate	Number of Deaths per 1000		
Deforestation	Removal or cutting of trees		
Dehydration	When the body doesn't have enough water	**Discrim-ination**	The unjust or prejudiced treatment of different groups of people - e.g. treating people differently because of their race or gender
Demographic Transition Model (DTM)	The transition from high birth and death rates to low birth and death rates over time as a country develops		
		Dispersed Settlement	A settlement made up of isolate individual buildings separated by a few hundred metres with no focus
Demography	The study of Population		
Dense Population	When there are a lot of people living in an area (e.g. London)	**Distributaries**	Channels in a delta. Formed when sediment is deposited in river channels, meaning

201

INFORMATION

	the water has to find an alternate route	**Epicentre**	The point on the surface directly above the focus
Diurnal Temperature Range	The range of temperature during a day (difference between the min. and max. temperature)	**Equality**	Similarities
		Equatorial Climates	A tropical climate located between the equator and the tropics. Rainforests are typically found here
Diversity	Having different elements or qualities - this could be a blend of different cultures within a country	**Erosion**	The process of material being removed by erosional processes by wind or running water
Domestic Tourist	A tourist who goes on holiday within their home country	**Erupting Volcano**	A volcano that is currently releasing lava or ash in a volcanic eruption
Dormant Volcano	A volcano that hasn't recently erupted but may erupt again in the future	**Euro (€)**	The common official currency of many EU member states
Drainage Basin	The area of land from which the water content is drained into a river or its tributaries	**Evaporation**	When liquid water in surface stores becomes a gas
		Evapo-transpiration	When water is transferred to the atmosphere by evaporation and transpiration
Dredging	The process of making a river channel deeper		
Drought	A prolonged period of low precipitation	**Even Population Distribution**	When an area's population is distributed regularly
Earthquake	Shaking or vibration of the earth's crust as a direct result from friction between 2 tectonic plates along a fault	**Exploitation**	Treating someone unfairly to benefit from them - e.g. people traffickers making a profit by charging high fees to take desperate migrants to their destination
Economic Development	The process of a country improving its citizens' socioeconomic wellbeing		
Economic Growth	The increase in amount of goods, services and profit produced over a period of time	**Extensive Farms**	Farms that use less or smaller inputs and less land
		Extinct Volcano	A volcano that **is** unlikely to erupt again as no magma is rising through it any more
Economic Migrants	Migrants who choose to migrate for money (e.g. job)		
Economically Active People	People who currently work, earn money and pay taxes	**Famine**	When there is not enough food for a given population
Ecosystem	A complex system of plants and animals that depend on each other and the environment	**Favelas**	Informal Settlements (in Brazil)
		Feedback	The profit being reinvested as an input
		Fertile (Soil)	Able to grow crops
Ecotourism	Tourism with an aim to have a low impact on the environment	**Flood Plain**	The area either side of the channel which floods when the river bursts
Emigrant	A person who leaves a country to live permanently in another country	**Flood Walls**	Concrete structures that prevent water from flooding an area
Enhanced Global Warming	When world temperatures rise as a result of human activity		

GLOSSARY

Flooding	When an area is submerged and covered with water due to a river's discharge being greater than its capacity
Focus	The point where an earthquake originates from
Freedom of Trade	When goods can move freely without tariffs or restriction
Gabions	Metal rock cages that absorb moving water's energy
Genetic Modification (GM)	When the genetic makeup of a crop has been artificially altered
Globalisation	The process by which the world is becoming more interconnected due to increased trade, with more cultural and economic exchanges
GNP per Capita	Gross National Product per Capita. A measure of the wealth or money generated by a country
Government Policies	A plan or set of regulations that is meant to influence and invoke change
Greenfield Land	Land that was previously farmland or was undeveloped
Greenhouse Gases	Gases in the atmosphere that prevent infrared radiation from escaping through the atmosphere. They keep the Earth at the ideal temperature for life when in constant, natural concentrations
Groundwater Flow	The movement of water through saturated ground
Groundwater Storage	Water stored in saturated ground. It makes the ground saturated, forming the water table
Groundwater Table	The boundary between saturated and unsaturated ground. It is the top layer of groundwater storage
Hadley Cells	The 2 atmospheric cells that straddle the Equator - they are between the Equator and the tropics. They are responsible for Equatorial and Hot Desert climates
Hamlet	A small settlement, smaller than a village
Hard Engineering Flood Prevention	Expensive, permanent structures that affect the flow of rivers, preventing flooding
Headland	A section of land jutting out into the sea
High Order Goods	Goods that people buy less frequently and compare for. They are sold at high order services
High Order Services	Services that sell high order goods with a large range and threshold population
High Order Settlements	Settlements providing lots of services (including high order services) with a large sphere of influence
High Technology Industries	Industries that involve the application of technology into a product
High Tech Industry	Industry producing the most advanced technology that is currently available - usually electronics
HIV	Human Immunodeficiency Virus - interferes with the ability to fight against disease. More common in less developed countries
Hot Deserts	An arid area of land with hostile conditions for life
Hot Spots	Parts of the mantle that are hotter than other parts
Human Development Index (HDI)	A composite index that blends together 3 of the more important development indicators [Adult Literacy, Life Expectancy and GDP per Capita]. It was designed by the United Nations to provide a uniform system of determining the level of a country's development. Because it is standardised, it allows analysts to compare

INFORMATION

	the development of different countries	Intensive Farms	Farms that use large amount of money, machines, technology and workers
Humanitarian Crisis	An event or events that threatens health, life and safety of a group of people	Interception	When an object (like a tree) stops the precipitation from reaching the ground
Humidity	The amount of water vapour in a given volume of air	Internal Migration	Migration within the country
Humus	Decaying organic matter	Internally Displaced	When people are forced to domestically migrate
Hydraulic Action	Erosion caused by the force and impact of flowing water, removing material	International Migration	Migration from one country to another
Hydrological Cycle (The)	The movement of water through the land, hydrosphere and atmosphere	International Tourist	A tourist who goes on holiday to another country
		Involuntary Migration	When Migrants are forced to migrate
Hydrosphere	All the water on the earth's surface	Isolated Settlement (or Dwelling)	A single building
Hygrometer	An instrument used to measure relative humidity. It is a combination of a wet and dry bulb thermometer	Land Use	The purpose that the land in the city is used for - this could be residential, industrial, open space or commercial
Immigrant	A person who lives permanently in a foreign country	Land Use Models	Theories trying to describe and explain the structure of urban city areas
Inequality	Differences	Lateral Erosion	Sideways Erosion
Infant Mortality	The number of new-born deaths under the age of 1 per 1000 live births	Lateral Motion	Side to side motion
Infiltration	When water travels from surface storage into the ground	Lava	Liquid molten rock above the Earth's surface
		LEDCs	Less Economically Developed Countries
Informal Settlements	Settlements built quickly without planning, often illegally, that ore more common in LEDCs. They are usually located on the Rural-Urban fringe	Levées	Natural ridges at the sides of a river channel
		Life Expectancy	The average age that someone is expected to live in a country
Inner City	The historic section of the town - often where industry used to be - mainly consisting of terraced houses	Linear Settlement	Settlements that form a line or arc shape
		Load	Eroded materials transported by a river or other water body
Inner Core	The hottest part of the earth, located at the very centre	Long Profile (of a river)	How a river's gradient changes over the course of a river
Inner Suburbs	Middle class residential built to house a growing population - generally consisting of semi-detached houses		
Inputs	Things that go into a system		

GLOSSARY

Longshore Drift	The movement of material in a zig zag motion along a coast by constructive waves
Low Order Goods	Goods that are frequently purchased, being very cheap. They are sold at low order services that are located conveniently for the consumer
Low Order Services	Services that sell low order goods with a local range and small threshold population
Low Order Settlements	Settlements providing few service - which are all low order services. It has a small sphere of influence
Magma	Liquid molten rock below the Earth's surface
Magnitude (Earthquake)	How strong the earthquake is (total amount of energy released) or the scale of its effect on the environment and people
Malnutrition	A lack of food or nutrition
Mangrove Swamp	A group of tropical trees growing in swampy areas near the sea
Mantle	Semi-molten part of the earth where convection currents flow. Magma, material in the mantle, forms the Earth's crust at constructive plate boundaries
Marine	Of the sea
Marram Grass	A type of grass that grows on newly-formed dunes. Its long roots help to bind the sand together
Meander Bend	A landform where a river moves from side to side across its flood plain as a result of erosion and deposition on each river bank
MEDCs	More Economically Developed Countries
Mercalli Intensity Scale	A method of measuring earthquake intensity based on its effects and its damage

Middle Order Goods	Goods that are purchased now and again - they are more expensive than low order goods, but still cheap compared to high order goods. They are sold at middle order services
Middle Order Services	Services that sell middle order goods with a medium range and threshold population
Middle Order Settlements	Settlements providing some services (including middle order services) with a moderately large sphere of influence
Migrant	A person who moves from one place to another
Migration	The movement of people
Mixed Farms	Farms that grow crops and rear animals
Morphology	Patterns or shape of something
Motorways	Major roads with a high traffic capacity. Here, cars move at high speeds in multiple lanes for long distance regional travel. You can find them in the United Kingdom (UK)
Mouth	Where a river meets the sea
National Park	An area of land protected by the government for people's enjoyment and wildlife preservation
Natural Decrease	Population shrinks as death rate is higher than birth rate
Natural Increase	Population grows as birth rate is higher than death rate
Natural Population Change	Changes in population as a result of birth and deaths
Net Migration Rate	The difference between the number of immigrants and emigrants throughout the year
NICs	Newly industrialised Countries

205

INFORMATION

Noise Pollution	Noise from industry or vehicles		bend loop is cut off by the river cutting through the neck of land during a flood
Non-Renewable Resource	Resources that cannot be replaced, or are replaced very slowly	**Package Holiday**	A bundle of flights and other services
Nucleated Settlement	Buildings in the settlement are built close to each other around a common centre	**Pastoral Farms**	Farms that rear animals
Oasis (pl. Oases)	A fertile area in a desert around a spring, pond, lake or other water source	**People Traffickers**	A person engaged in illegal movement of people from one place to another (e.g. migrants from Syria to Europe)
Oceanic Crust	Less dense tectonic plates than continental crust where most of the planet's oceans are located	**Permanent Migration**	People who migrate and don't return
		Persecution	When someone is prosecuted or attacked for a racist reason
Open System	A system which allows energy and substances to enter and leave	**Plate Boundaries**	Where 2 plates meet. Volcanoes and Earthquakes occur here
Optimum Population	When there are enough resources to support the population to a good standard of living	**Population**	The inhabitants of an area (e.g. a country)
		Population Density	The number of people living in an area
Origin Country	The country which a person is travelling from	**Population Distribution**	How the population is distributed in an area
Outer Core	The part of the earth surrounding the inner core, located beneath the mantle	**Population Growth Rate**	Natural Increase / 10
Outer Suburbs	Upper class residential that is more modern - mainly consisting of detached houses with driveways and gardens	**Population Pyramids**	Diagrams used to show age and sex structure in a country
		Population Structure	The composition of a population. Most frequently shown using a population pyramid diagram
Outputs	Things that come out of a system	**Precipitation**	Water falling in any form from the sky
Overcrowding	More people living in a city than the city was designed to hold - meaning that strain is put on public services	**Prediction**	Forecasting what will happen in the future
		Pressure (Air)	The amount of weight exerted on the Earth's surface in a column of atmosphere
Over-cultivated	When the land has been farmed, cultivated and used excessively to the extent where it is not as fertile as it used to be	**Prevailing Wind**	The most common wind direction in an area
		Primary Sector	Industries that involve the collection of raw materials
Overland Flow	When water travels across the surface of the ground	**Processes**	Things that happen to inputs to turn them into outputs
Over-population	Where there are too many people to be supported to a good standard of living by the resources of the country		
Oxbow Lake	A curved lake which is formed when a meander		

GLOSSARY

Producers	Plants that capture the sun's energy, storing it as glucose or starch
Pro-Natalist Policy	Aims to increase population growth by encouraging couples to have more children
Public Services	A service provided by the government
Pull Factors	Encourage a person to go to a place
Push Factors	Push a person to leave a place
Quality of Life Index	A composite index that uses GDP, Life Expectancy, Happiness and how people feel about their family, workplace and local community. A country with a higher value is happier, and often more developed, than a country with a lower value
Quaternary Sector	Industries that involve knowledge based jobs
Rain Gauge	A device which measures the amount of precipitation by collecting rainfall
Rain Shadow Effect	When wind goes up a mountain, the moisture condenses into a cloud - meaning that the area downwind of the mountain has dry air
Rainforest Canopy	The upper layer of a rainforest where most of the life is located
Rainwater Harvesting	The collection and storage of rainwater in small collection vessels or in massive reservoirs
Range	The distance people are willing to travel to buy a product or service
Reforest-ation	Replacement of trees that have been cut down
Refugees	Migrants who migrate to escape from an event
Relative Humidity	The amount of water vapour in the air as a percentage of the amount of water vapour needed for the air to be saturated
Relief	The shape of the land
Renewable Resources	Resources that can be replaced
Reservoirs	Pools of water that build up behind dammed rivers
Resources	A valued or needed material
Richter Scale	A method of measuring earthquake intensity by the magnitude of the earthquake
River Cross-Sectional Area	The area of a slice of flowing water
River Discharge	When the water carried in a river is released into the sea
Rural Depopulation	The decreasing population of rural areas as people move to urban areas
Rural-Urban Fringe	The area between the edge of the outer suburbs and the countryside
Rural-Urban Migration	Migration from a rural environment (e.g. a village) to an urban environment (e.g. a city)
Saltation	Small pebbles and stones are transported (e.g. by bouncing) along the bottom of the water body
Sand Dunes	A hill of sand that builds up from deposition processes
Sanitation	Public health conditions - comprised of clean drinking water and proper sewage disposal
Saturated Air	When the relative humidity is 100%
Saturated Ground	Rock or soil where all the pores are filled with water because they are beneath the groundwater table
Secondary Cone	Another vent for the volcano to erupt through. Forms when magma leaks through the alternating layers of lava and ash
Secondary Sector	Industries that involve manufacturing products

207

INFORMATION

Seismic Design	Reducing the impact of earthquakes by building design. It aims to prevent buildings from collapsing in an earthquake by allowing them to absorb the earthquake's energy or being strong enough to stay standing
Services	Facilities that are offered to people. They have threshold populations
Settlement	A place where people live. It can be of any size, from a small house to a World City
Settlement Hierarchy	A way of putting settlements in order of importance
Severe Flooding	Large scale flood events that create massive damage to the environment and settlements
Shield Volcanoes	Volcanoes with gentle slopes and a wide base. Formed when the lava is more basic.
Shingle	Rounded beach material between the size of boulders and sand
Silica Content	The amount of silica in Lava. A higher silica content means the lava is more acid - with a higher viscosity and is slower-moving. A lower silica content means that the lava is more basic - with a lower viscosity and increased fluidity
Site	The land which a settlement or factory is built on
Situation	The position of the settlement in relation to the surrounding area
Slash and Burn	A farming technique that involves the felling and burning of trees to create new fields
Slums	Informal Settlements
Soil Conservation	The process of reducing soil erosion
Soil Erosion	The removal of soil from the ground by wind or running water

Soil Moisture Storage	Water stored in pores in unsaturated ground
Solution	Minerals are dissolved in the water
Source	The start of a river
Sparse Population	When there are few people living in an area (e.g. Sahara Desert)
Sphere of Influence	The area that the service serves
Spit	A long, low and narrow ridge of sand or shingle that has one end attached to the coast, with the other end in the sea. They are formed by coastal deposition
Squatter Settlements	Informal Settlements without legality
Staple Crops	The most common foods eaten in an area
Stevenson Screen	A special wooden box designed to shield meteorological instruments from the environment, allowing them to take more accurate readings
Strata	Layers of substances - for instance of different types of volcanic material, rock or soil
Stratovolcanoes	Volcanoes made up of alternating strata of lava and ash. Formed when the lava is more acid
Subduction	The process by which a denser plate becomes magma as it is forced under another, less dense plate
Subduction Zone	Where a denser oceanic plate is forced to plunge below a much thicker continental plate. As the plate descends, it melts, becoming magma
Subsistence Farming	Farming that produces food for the farmer and their family
Sunshine Hours	The number of hours in a day where there is sunlight
Sunshine Recorder	A device that records the sunshine hours using a light sensor and a storage

GLOSSARY

	mechanism to record the number of hours of sun there were in a day	Throughflow	The movement of water through unsaturated ground
Surface Storage	Water held on the surface of the ground	Tourism	The industry that provides services to tourists
Surface Water	Water from rivers and lakes	Town	A settlement larger than a village but smaller than a city - it has a name, a boundary and usually a local government or council
Suspension	Light materials are carried near the surface of the water, potentially tinting the water body. When soil is carried, a murky brown colour is created	Traction	Large, heavy boulders and rocks are rolled along the bottom of the water body
Sustainability	Fulfilling our resource requirements without hindering the access to these benefits for future generations. This means that we should only use the Earth's resources at the rate that they can be replenished	Traffic	Vehicles moving on a road
		Traffic Jam	A line of built up traffic that is stationary or near stationary
		Transition Zone	Where factories and industry are located
		Transnational Corporations (TNCs)	Corporations that operate, manufacture and trade all over the world. They have factories and offices in many different countries
Sustainable Tourism	Activities or services that are sustainable in a social, economic and environmental sense	Transport-ation	When liquid water evaporates from vegetation
Swash	The forwards motion of a wave	Tributary	A smaller river or stream that flows into a larger river or lake
Tax Payers	People who pay taxes	Tropical Rainforests	A forest rich in life that forms near the equator in equatorial climates
Taxes	A compulsory contribution of money to the government		
Tectonic Plates	Large, interlocking slabs of crust that are moved across the Earth's surface by convection currents	Tropics	About 23 degrees north or south of the equator
		Tsunami	A high wave caused by an earthquake event
Temperature	The degree of warmth in the air	Under-population	When countries have insufficient workers to exploit their resources efficiently, to support retired populations and to provide growth
Temporary Migration	Migrating for a short or set period		
Tertiary Sector	Industries that involve providing a service		
The European Union (EU)	A political and economic group of member countries in Europe that act in as a single block	Uneven Population Distribution	When some areas have lots of people and others very few
		Unsaturated Ground	Rock or soil above the groundwater table. The soil and rock may contain air in their pores
Thermometer	An instrument used to measure temperature		
Threshold Population	The minimum number of people needed for a service to be offered or to be available	Urban Decay	When parts of a city become run-down and undesirable to live in

209

INFORMATION

Urban Growth	The expansion of towns and cities - due to more buildings being built. This means the urban area covers more land and can support a larger population	**Voluntary Migration**	When Migrants choose to migrate
Urban Regeneration	The process of improving an urban environment	**Waste**	The unwanted things that come out of a system
Urban Sprawl	When the city expands into the countryside	**Water Deficit**	When there is less water than needed
Urbanisation	The increase in the proportion of people living in towns or cities in an area	**Water Mismanagement**	When water sources are overused
US Dollar ($)	The currency used in the United States of America (USA). GDP is measured in US Dollars	**Water Pollution**	Contamination of a water supply
		Water Shortages	When water isn't available or isn't available in sufficient quantities to fulfil the country's needs
Vertical Erosion	Downwards Erosion	**Water Surplus**	When there is more water than needed
Vertical Motion	Up and down motion	**Watershed**	The edge of the drainage basin. It is also the boundary between 2 different drainage basins
Village	A group of buildings in a rural area - larger than a hamlet and smaller than a town	**Wave-Cut Platforms**	Narrow, flat areas that are found at the bottom of a cliff. Created by the wave's erosion and the collapse of the ceiling of a wave-cut notch
Viscosity	How thick something is. Low silica content in the lava results in a low viscosity, which means the lava is runnier	**Weather Elements**	The parts of Weather
		Weather	The state of the atmosphere at a certain time
Visual Pollution	Impacting the beauty and appearance of the natural world — e.g. quarries — or something that spoils the beauty of the landscape	**Weathering**	The process of something being eroded by exposure to air
		Wind	Air moving over the surface of the ground
Volcanic Bombs	Rocks thrown out of the volcano as part of the eruption	**Wind Gusts**	Fluctuations in wind speed
		Workers	People who have a job, work, earn money and pay taxes
Volcanic Eruptions	A violent release of volcanic material and steam above the surface from below the surface	**Workforce**	People available or currently in work in a country
		Working Population	The number of people who are at working age
Volcano	A place in the Earth's crust where magma from the mantle is allowed to reach the surface (through a vent)	**World City (Global City)**	A city that plays an important role in the global economy - often serving both residents and visitors. It has a global sphere of influence
Volcanologists	People who study volcanoes. They predict volcanic eruptions by making observations and forecasts		

GLOSSARY

Yield (Crop Yield)	The amount of crop harvested from an area of land. A high yield is important for farmers, as it creates more profit.
Youth Population	Young people who are dependent - the UN defines this as people between the ages of 15 and 24

INFORMATION

COMPLETE CONTENTS

How to Use This Guide 2
Theme 1: Population and Settlement 4
Population 5
   1.1 Population dynamics 6
      Reasons for a rapid increase in the world's
      population 6
      Overpopulation and Underpopulation 8
         Problems Caused 8
      Main causes of a change in population size 8
      Contrasting Rates of natural population change 9
         Demographic Transition Model 10
      Population Policies – China's 1 Child Policy 12
         Impacts 12
         Future of the policy 13
      Case Studies 14
         Overpopulation – Bangladesh 14
         Underpopulation – Australia 15
         High rate of natural population growth –
         Bangladesh 16
         Declining population – Japan 17
   1.2 Migration 18
      Key Words 18
      Reasons for Population Migration 18
      Impacts of Migration 19
      Case Study – Refugee International Migration to
      Europe 20
   1.3 Population Structure 22
      Different Population Groups 22
      Population Structures and Economic Development
      24
         LEDCs 24
         MEDCs 24
      Case Study - Japan 25
   1.4 Population Density and Distribution 27
      Case Studies 28
         Densely Populated – Greater London Area,
         UK 28
         Sparsely Populated – Himalaya Mountains,
         Asia 29
Settlement 31
   1.5 Settlements and Service Provision 32
      Patterns of Settlement 32
      Factors influencing settlements 34
      Hierarchy of Settlements and Services 35
         Settlement Hierarchy 35
         Services 36
      Case Study – Isle of Man 38
      Case Study – Greater Manchester, United
      Kingdom 39
   1.6 Urban Settlements 40
      Land Use 40
         Characteristics of Land Use 40
         Changes in Land Use 42

      Effect of rapid urban growth 42
      Problems of Urban Areas 43
      Urban Decay 45
      Case Study – Manchester 46
   1.7 Urbanisation 48
      Rapid Urban Growth 48
      Impacts and Solutions 49
         LEDCs – Squatter Settlements 51
      Case Study – Mumbai, India 53
Theme 2: The Natural Environment 56
   2.1 Earthquakes and Volcanoes 57
      Tectonics 58
      Volcanoes 62
         Types and Causes of Volcanoes 62
         Features of Volcanoes 64
         Effects of Volcanic Eruptions 65
         Hazards and Opportunities 66
         Reducing Impacts 67
      Case Study – Eyjafjallajökull, Iceland 68
      Earthquakes 70
         Features of Earthquakes 70
         Causes of Earthquakes 70
         Hazards of Earthquakes 71
         Reducing Impacts 72
      Case Study – Kobe Earthquake, Japan 72
   2.2 Rivers 75
      Drainage Basins 76
         Characteristics 77
         Processes 77
      River Characteristics 78
      River Processes 79
      River Landforms 80
         Upper Course 81
         Waterfalls 81
         Middle Course 82
         Lower Course 84
         Cross Sections + Summary 85
      Hazards and Opportunities 85
      Reducing the Impacts 86
      Case Study – Mekong River, SE Asia 88
   2.3 Coasts 91
      Coastal Processes 93
      Coastal Landforms 95
         Coastal Erosion 95
         Coastal Deposition 97
      Coral Reefs 99
      Mangrove Swamps 101
      Hazards and Opportunities 102
      Managing Impacts of Coastal Erosion 103
      Case Study – New Forest Coastline, Hampshire,
      UK 104
   2.4 Weather 105
      Elements of Weather and their Collection 106

COMPLETE CONTENTS

Precipitation - Rain Gauge 106
Temperature - Thermometers..................... 107
Humidity - Hygrometers 107
Sunshine - Sunshine Recorder 108
Pressure – Barometers 109
Wind Speed - Anemometers 110
Wind Direction - Wind Vanes 110
Clouds... 111
Digital Instruments 112
Stevenson Screens 112
Calculations + Interpreting Graphs and
Diagrams 113
2.5 Climate and Natural Vegetation 115
Characteristics 117
Equatorial Climates 117
Factors affecting characteristics 118
Hot Desert Climates 119
Factors affecting characteristics 120
Climatic Graphs 120
Ecosystems 122
Tropical Rainforest 122
Hot Desert... 125
Deforestation of Tropical Rainforest 127
Causes 127
Effects 128
Why Protect the Rainforests? 128
How? 129
Case Study – Tropical Rainforest – Bornean
Rainforest 129
Case Study – Hot Desert – Namib Desert,
Namibia 131
Theme 3: Economic Development 134
3.1 Development 135
Development Indicators................................. 136
Inequalities 137
Industrial Sectors.. 138
Globalisation 139
The process of Globalisation 139
Impacts of Globalisation 141
Case Study – Nike 142
3.2 Food Production 143
Types of Farming 144
Agricultural Systems.. 144
Food Shortages 147
Causes 147
Effects 147
Solutions 148

Case Study – Rice Farming, Bangladesh 149
Case Study – Famine, Horn of Africa 150
3.3 Industry 151
Industrial Systems.. 152
Distribution and Location 153
Types of Industry 154
Case Study – Toyota, Burnaston, UK 156
3.4 Tourism 159
Growth of Tourism 160
Attractions of The Physical and Human
Landscape 162
Benefits and Disadvantages 163
Sustainable Management................................ 164
Case Study – Iceland.. 165
3.5 Energy 167
Growing Consumption 168
Non-Renewable Fossil Fuels 169
Renewable 170
Fuelwood 172
Nuclear Power – The Future?......................... 172
Case Study – Iceland.. 174
3.6 Water 175
Water Supply 176
Water Use... 177
Surplus and Deficit 178
Water Shortages and Management 178
Why are there Water Shortages? 178
The Impact of Water Shortages 180
Managing Water.. 180
Case Study – Lesotho 181
3.7 Environmental Risks of Economic Development
.... 183
Threats to the Natural Environment 184
Why is there a threat?................................... 184
Soil Erosion 184
Desertification... 186
Enhanced Global Warming 187
Pollution 189
Being Sustainable 189
Resource Conservation 191
Case Study – The Aral Sea, Kazakhstan and
Uzbekistan 192
Licences for Media Used 195
Acknowledgments 196
Command Words.. 198
Glossary 199
Complete Contents... 212

213

Printed in Great Britain
by Amazon